T0073620

WHEN GALAXIES WERE BORN

WHEN GALAXIES WERE BORN

THE QUEST FOR COSMIC DAWN

RICHARD S. ELLIS

PRINCETON UNIVERSITY PRESS

PRINCETON AND OXFORD

Published by Princeton University Press
41 William Street, Princeton, New Jersey 08540
99 Banbury Road, Oxford OX2 6JX

press.princeton.edu

Library of Congress Cataloging-in-Publication Data

Names: Ellis, Richard S. (Richard Salisbury), 1950– author.
Title: When galaxies were born : the quest for cosmic dawn / Richard S. Ellis.
Description: Princeton, New Jersey : Princeton University Press, [2022] |
    Includes bibliographical references and index.
Identifiers: LCCN 2022002210 (print) | LCCN 2022002211 (ebook) |
    ISBN 9780691211305 (hardback) | ISBN 9780691241678 (ebook)
Subjects: LCSH: Galaxies—Formation. | Cosmology. | Astronomy—Observations. |
    BISAC: SCIENCE / Space Science / General | SCIENCE / Space Science / Astronomy
Classification: LCC QB857 .E45 2022 (print) | LCC QB857 (ebook) |
    DDC 523.1/12—dc23/eng20220524
LC record available at https://lccn.loc.gov/2022002210
LC ebook record available at https://lccn.loc.gov/2022002211

British Library Cataloging-in-Publication Data is available

Editorial: Ingrid Gnerlich and Whitney Rauenhorst
Production Editorial: Kathleen Cioffi
Text and Jacket Design: Karl Spurzem
Production: Jacqueline Poirier
Publicity: Sara Henning-Stout and Kate Farquhar-Thomson
Copyeditor: Maia Vaswani

Jacket images: (Top) Hubble views a cosmic Interaction, NASA.
(Bottom) Mauna Kea Observatories by Julian Abrams

This book has been composed in Arno Pro with Trade Gothic Next

Printed on acid-free paper. ∞

Printed in the United States of America

10 9 8 7 6 5 4 3 2 1

CONTENTS

ILLUSTRATIONS

Figures

Colour Plates (following page 28)

Colour Plates (following page 108)

PREFACE

For more than a century the idea that we might travel back in time has been a staple of science fiction and popular culture. But astronomers *can* time-travel! Because of the fixed speed of light, when a large telescope views a faint galaxy at a great distance, we look back to an earlier time when the light left that galaxy. In this way, astronomers can "time-slice" the universe and reconstruct its evolution. Galaxies are the visible fabric of the universe, immense assemblies of stars, gas and dust particles, and a major drive in observational astronomy since the 1970s has been to probe the universe back to the time when the earliest examples first formed. This moment, called "cosmic dawn," represents the time when the universe was first bathed in starlight. As we are, ourselves, formed of atoms synthesised in generations of stars over billions of years, the quest for cosmic dawn can truly be regarded as a search into our own origins and place in the cosmos.

This book tells the story of finding and studying ever-more-distant, and hence earlier, galaxies from a personal perspective, as I have now witnessed the transformation of observational astronomy over a career of five decades. Progress has been driven by technological advances that have led to larger and more powerful telescopes and more sensitive instruments. But the ambition and creativity of astronomers, graduate students, and postdoctoral researchers has also been an important factor, all driven by the thrill of discovering the unknown. In the relentless competition to find and study ever-more-distant sources, there have been spectacular achievements and disappointments, as is typical of the fast-paced world of modern science.

At the heart of the story is the telescope itself, a sophisticated machine that enables us to achieve the seemingly impossible: to piece

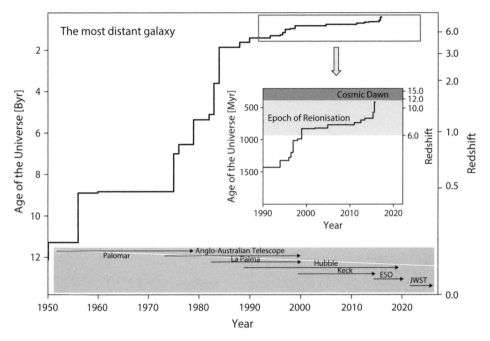

FIGURE 0.1. The most distant known galaxy versus publication date in terms of the age of the universe when the galaxy is being observed (left axis) and its redshift (see chapter 2; right axis). I also indicate below the period in my story during which the contribution of each observatory is discussed. Byr, billion years; Myr, million years.

together a coherent view of how our universe developed from darkness into its impressive starlit beauty today. The famous cosmologist and popular writer Sir Fred Hoyle once wrote, "a large telescope is a good example of the things which our civilisation does well."[1] Engineers, opticians, astronomers, and fundraisers work together, in some cases over several decades, to construct and assemble these temples of science at remote sites, far from the light pollution of urban areas where their components were manufactured. Because of the dramatic locations of these observatories, modern astronomy also has a romantic element. While most research scientists work in laboratories close to home, the observational astronomer can stand on a mountain summit far away, witnessing

1. *The Creation of the Anglo-Australian Observatory*, S.C.B. Gascoigne, K.M. Proust, and M.O. Robins (Cambridge University Press 1990), p282.

a dramatic sunset and the onset of darkness before entering the telescope enclosure in anticipation of making new and profound discoveries about the universe.

My story develops these themes for each observatory in turn, from the pioneers who realised the prospects and carefully selected their mountain sites to the visionary leaders who supported these endeavours and, later, to the astronomers who made the discoveries. As a result, the account is not presented strictly chronologically because, as in all science, progress occurs in parallel through the use of varying techniques in different places. Hopefully the timeline in figure 0.1 will help guide the reader as the story unfolds. Furthermore, it is a personal account rather than an all-inclusive academic review of the history of our understanding of cosmic evolution. My hope is that such an "insider view" will be more revealing to a general reader of how this remarkable progress in probing the early universe has unfurled. It is possible, of course, that others will have a different recollection of the events. I apologise for any potential errors or omissions, which are entirely my responsibility.

The idea for this book followed a request made by a professional photographer, Julian Abrams, who simply asked me if I could arrange a tour of large telescopes for a photographic exhibition he wished to assemble. With generous funding from a former astronomer and colleague, Paul Atherton, we toured nearly all of the telescopes I have used in my career, and as a result many of Julian's striking photographs illustrate my story. I thank Julian for his boundless enthusiasm as a travel partner and Adrian Hall for helping with the complex logistics.

I have been fortunate, during my career in seven academic institutions and observatories in three countries, to have worked with numerous talented researchers, many of whom feature in the story. I particularly thank Steve Phillipps, Tom Shanks, Bruce Peterson, Tom Broadhurst, Matthew Colless, Karl Glazebrook, Bob Abraham, Andy Bunker, Jean-Paul Kneib, Mike Santos, Johan Richard, Dan Stark, Tucker Jones, Matt Schenker, Adi Zitrin, Guido Roberts-Borsani, Nicolas Laporte, Koki Kakiichi, Sarah Bosman, Romain Meyer, and Harley Katz, without whose dedication and enthusiasm most of the progress we have made would never have occurred. Likewise, I have worked with many talented instrumentalists who built innovative spectrographs for our work, including Peter Gray, Ian Parry, Keith Taylor, Paul Atherton, Sue Worswick, and Jeremy Allington-Smith.

In considering the above list of collaborators, it will doubtless be striking to today's readers that most of the players in my story (fortunately not all) are men. Given the timeline of over five decades, this reflects the sad fact that, until quite recently, few women were able to build careers through studies of distant galaxies. Although the situation is slowly improving, it is clear that we must continue to encourage women and under-represented groups to contribute to our understanding of the universe.

Heidi Aspaturian at Caltech is thanked for her editorial comments and meticulous reading of early drafts of this book. Her patience with my desire to complete this story is much appreciated. I likewise thank Ian Corbett, Roger Davies, Ofer Lahav, Patrick McCray, and John Peacock for their insights into some of the early history. Kate Whitaker and two anonymous readers are thanked for very valuable comments on an earlier draft of my manuscript. I owe enormous gratitude to Ingrid Gnerlich at Princeton University Press for her prolonged support, advice and encouragement, and to both her and her associates Whitney Rauenhorst and Kathleen Cioffi for their patience with me during various stages of the publication process. I also thank the Press's copyeditor, Maia Vaswani, for her work.

Like most observational astronomers, I have necessarily spent countless weeks far from home, thereby missing several Christmases, wedding anniversaries, and family birthdays. I thank my patient wife Barbara and my children Hilary and Tom for their willingness over the years to allow me to do what I enjoy the most—staying up all night observing with a large telescope.

Richard Ellis
Cambridge, August 2021

ABBREVIATIONS

2dF	Two-degree field facility (AAT wide-field multi-object spectrograph and optics)
AAO	Australian Astronomical Observatory (formerly Anglo-Australian Observatory)
AAT	Anglo-Australian Telescope
AGN	active galactic nucleus
ALMA	Atacama Large Millimeter Array
APM	Automated Plate Measuring machine at Cambridge University
AURA	Association of Universities for Research in Astronomy
BAA	British Astronomical Association (amateur organisation)
CANDELS	Cosmic Assembly Near-Infrared Deep Extragalactic Survey (taken with HST)
CCD	charge-coupled device (digital detector)
CERN	Centre for European Nuclear Research
CFHT	Canada-France-Hawaii Telescope on Maunakea
CFRS	Canada-France Redshift Survey (redshift survey undertaken with CFHT)
DEIMOS	Deep Imaging Multi-Object Spectrograph (a Keck instrument)
EFOSC	ESO Faint Object Spectrograph and Camera (instrument at La Silla)
ELT	ESO's Extremely Large Telescope at Cerro Armazones
ESA	European Space Agency
ESO	European Southern Observatory
FOS	Faint Object Spectrograph (two versions, both used on La Palma)
GTC	Gran Telescopio Canarias (a.k.a. GranTeCan—a Spanish telescope on La Palma)
HDF	Hubble Deep Field (a deep image secured with HST)
HST	Hubble Space Telescope
HUDF	Hubble Ultra-Deep Field (a deeper image secured with HST)
HUDF12	the 2012 Ultra-Deep Field (an augmented version of HUDF)
IAC	Instituto de Astrofísica de Canarias (which manages the La Palma observatory)
INT	Isaac Newton Telescope
IPCS	Image Photon Counting System (electronic detector)
JWST	James Webb Space Telescope
LDSS	Low Dispersion Survey Spectrograph (two versions, one at AAT, one at WHT)
LST	large space telescope (later descoped and became HST)
LTP	SERC's Large Telescope Panel
MDS	Medium Deep Survey (undertaken with HST)

MIRI	Mid Infrared Imager (an instrument on board JWST)
MOSFIRE	Multi-Object Spectrograph for Near-Infrared Exploration (a Keck instrument)
MSSSO	Mount Stromlo and Siding Spring Observatories
MX	an automated fibre positioner developed at the University of Arizona
NGST	Next Generation Space Telescope (later named JWST)
NIRCam	Near Infrared Camera (an instrument on board JWST)
NIRISS	Near Infrared Imager and Slitless Spectrograph (an instrument on board JWST)
NIRSpec	Near Infrared Spectrograph (an instrument on board JWST)
NIRSPEC	Near Infrared Spectrograph (a Keck instrument)
NNTT	National New Technology Telescope (a US proposal for a large telescope)
NOAO	National Optical Astronomy Observatory (United States)
NSF	National Science Foundation (US funding agency)
RAS	Royal Astronomical Society (United Kingdom)
RGO	Royal Greenwich Observatory
SERC	Science and Engineering Research Council (UK funding agency)
SIRTF	Space Infrared Telescope Facility (later named the Spitzer Space Telescope)
SKA	Square Kilometre Array (a future radio interferometer)
ST-ECF	Space Telescope—European Coordinating Facility
STScI	Space Telescope Science Institute (Baltimore, MD)
TAC	time allocation committee
UCL	University College London
UKIRT	UK Infrared Telescope (on Maunakea)
UKST	UK Schmidt Telescope (in Australia)
VLT	ESO's Very Large Telescope at Cerro Paranal
WFC	Wide Field Camera (an optical camera onboard HST)
WFC3	Wide Field Camera 3 (an ultraviolet/optical and near-infrared camera on HST)
WFPC2	Wide Field Planetary Camera (an optical camera which replaced WFC)
WHT	William Herschel Telescope
WMAP	Wilkinson Microwave Anisotropy Probe (satellite studying cosmic background)

WHEN GALAXIES WERE BORN

1

Out into Space

We live in a time full of remarkable astronomical discoveries. Scarcely a week goes by without media reports of some interesting new celestial object, be it an Earth-like planet orbiting a nearby star, an object of unknown origin arriving in the solar system or, as discussed in this book, the most distant known galaxy seen as it would have appeared when its light first set out for Earth billions of years ago. The universe fascinates many of us, and increasingly so as the pace of discovery accelerates. Unlike some other scientific disciplines, which require the understanding of difficult concepts with unfamiliar terminologies, astronomy has the advantage that everyone can understand the fascination of exploring outer space and discovering what's out there. Who hasn't, at one time or another, pondered such fundamental questions as, are we alone in the universe? Where did the world around us and the worlds beyond us come from? What does the future hold for the universe and our place within it? And what can we learn from gazing billions of years into its past?

My fascination with astronomy dates from childhood. When I was 6 years old, I visited the public library in the small coastal town of Colwyn Bay in North Wales, where I was born and grew up. One day I found a book in the children's section that set me on a career path of five decades as a professional astronomer. Exactly why I picked up this book is unclear to me now. It was a little blue book entitled *Out into Space*, with no striking illustrations on its cover or inside. It describes the fictional adventures of a young brother and sister who go to stay with their eccentric Uncle Richard (!), an astronomer with a telescope in his back garden. To this day I remember the chapter in which Uncle Richard persuades his niece and nephew to get up at six in the morning to observe the planet Mercury through his telescope. Gazing at Mercury, the children are fascinated to observe it as a small pinkish crescent, and they are struck by how remote it seems.

The book was written by (later Sir) Patrick Moore (1923–2012), Britain's most famous amateur astronomer and presenter of the BBC's *The Sky at Night*—a monthly documentary programme on astronomy. He presented the programme from 1957 until a posthumous broadcast in 2013, making it the longest-running series with the same host in television history. I was a guest on this programme twice in the 1990s, and, during my first appearance, I mentioned how Moore's little blue book had ignited my youthful interest in astronomy. To my surprise and delight, he subsequently sent me a signed version. It appears to have been his personal copy (this time with an illustrated cover). Rereading the book more than six decades after that trip to the library, I find it still evokes the childhood wonder of exploring the universe (see plate 1).

Moore's book set me on a course of reading everything I could find about astronomy. Public interest in the subject certainly grew after Soviet cosmonaut Yuri Gagarin became the first human to ride a rocket into outer space and make one full orbit of the Earth in 1961. I was in year six of primary school (the British equivalent of the American fifth grade) at the time, and I was asked by my teacher to describe the importance of this achievement to the class. By this time most of my classmates knew that I wanted to become an astronomer. The next logical step was to get hold of a telescope. Here I got a bit of help from the father of a friend. He generously gave me a small 4-inch reflecting mirror, which I then began to figure out how to fashion into a telescope. Acting on the advice of an older cousin, I managed to find a cardboard tube of approximately the right dimensions in a carpet shop and varnished it with a paint brush. With an eyepiece from a pair of marine binoculars that my seafaring father had acquired during his career as a captain in the British Merchant Navy, I now had the optical components. Then came the challenge of cutting the carboard tube to the right length, while keeping the mirror and eyepiece installed and pointing at the moon. This was more than a little nerve-wracking lest I cut the tube too short, in which case the telescope would never be in focus!

Finally I was ready to set up an observation station in the back garden. Even though I had a reasonably good idea of what to expect after all my reading, I was still unprepared for my first clear sight of the magnified night sky. I was immediately struck by the various colours of bright stars (indicative of their different temperatures), and, on subsequent nights, I followed the orbits of Jupiter's four largest moons and inspected craters and mountains on our own moon. These were exciting times for a young boy. I found I couldn't wait for it to get dark: there was so much to explore. But Wales is famous for its rainy climate, and, regardless, it is often overcast. The uncooperative weather, along with the interfering glare of nearby street lights and the limitations of my primitive telescope, which could not track the movement of stars across the sky, led to much frustration.

Today, youngsters with a keen interest in astronomy are likely to have more opportunities for encouragement and practical support. Numerous sites on the internet provide information about purchasing small telescopes and offer advice about assembling them; mobile-phone applications make it possible to view the night sky's appearance at any time and place across the globe. Local astronomical societies host viewing nights and offer talks and even workshops with both professional and experienced amateur astronomers.

North Wales in the early 1960s offered no such opportunities. My high school had no scientific societies, and even my parents, who generally kept close track of my educational progress and stressed the importance of academic success, rarely came outside to share my enthusiasm. Books from the public library and Patrick Moore's television programme were my only sources of information. Fortunately, Colwyn Bay's library had an excellent collection of quite advanced astronomy books, although my junior member's ticket wouldn't let me borrow books from the adult section. I had to be resourceful. From careful observations over several visits, I identified a librarian who appeared not to be aware of the difference between books in the children's and adults' sections. One afternoon I waited patiently until it was her turn at the desk and promptly presented her with a selection of astronomy books from the adult section to check out. This strategy worked well until one memorable day when, just as I marched up to the librarian in question, she was relieved of duty and replaced by another, who told me smugly, "You can't take out these books from the adult section on a junior ticket!" Despite that momentary setback, I eventually managed to extract and read most of the astronomy books in the adult section. Years later, I learned that the Colwyn Bay library had been established in 1904 with a benefaction of several thousand pounds from Andrew Carnegie, the wealthy Scottish American industrialist and philanthropist who played a major role in the development of Californian astronomy.[1]

Having exhausted the local library's resources, I began to look elsewhere for guidance and advice on how to develop my interest in astronomy. At about the age of 15, I joined the British Astronomical Association (BAA), which held regular meetings and organised amateur activities that involved coordinated telescope observations of the sun, planets, variable stars, comets, and so on. Unfortunately, all these events were based in or around London. It was impractical to hope to get involved at a distance of nearly 250 miles away in the wilds of North Wales. I did send for, and received, a BAA brochure entitled "Astronomy as a Career" (price 1 shilling) but found it painted a gloomy picture, warning of years of study and many hurdles to overcome before I could "enter the holy of holies: the dome of a large telescope for a night's observations."[2] Although I later found out that some of this advice was true (I didn't get to use one of the world's largest telescopes until more than a decade later), these were hardly words to encourage and inspire a 15-year-old.

The BAA brochure did offer me some practical and, as it turned out, highly useful advice. Mixed in with the "holy of holies" verbiage was the hard-headed admonition that modern astronomy is a rigorous and

1. *Colwyn Bay: Its Origin and Growth*, Norman Tucker (Colwyn Bay Borough Council 1953), p214.

2. *Astronomy as a Career*, E. A. Beet and R. H. Garstang (British Astronomical Association 1962), p5.

challenging *physical science*. To become a professional astronomer, a fascination with discovering the night sky, rewarding and exciting though that activity can be, is insufficient. A thorough grasp of mathematics and physics is essential. Indeed, in my career, I have met more than a few professional astronomers who are completely unfamiliar with the constellations, unable to name any bright star (other than the sun!), and generally content just to grapple with equations and program supercomputers.

In this regard I was fortunate in high school to have a dedicated physics teacher, Mr. E.O.P. Williams, who introduced me to the magic of applying the laws of physics across a wide range of everyday life. I was captivated by how familiar words such as "force," "energy," and "power" acquired tangible physical meaning, and by how physical laws could be used to predict the behaviour of objects in the real world. However, the headmaster of my high school was quite concerned when I told him I wanted to become an astronomer. A strict Welshman, unpopular even with his staff and willing to use corporal punishment on his students (including me), he tried to discourage me. He claimed he knew someone working at the Royal Greenwich Observatory who was "going nowhere fast." On the home front, my mother nurtured hopes that I would become a medical doctor or banker. Fortunately, she eventually recognised my unwavering determination and came around to accept that I was going to be an astronomer.

Teenage life was not, of course, all about academic study and preparations for a future career. The year was 1966, and a cultural revolution was under way in Britain, led by the Beatles and Rolling Stones, whose influence on music, fashion, and acceptable behaviour permeated every facet of adolescent life. Like teenage boys throughout Britain and the United States, my school friends and I formed a rock band called The Omegas and had some fun times performing at local venues. Not surprisingly, my burgeoning distaste for authority led to much friction with the headmaster, who had made known to all pupils his aversion to long hair. The school rule was that a boy's hair should not touch his collar. To this day I recall encountering the headmaster in the school corridor and rapidly assuming a posture with head bent forward and collar rolled back to be compliant. In my final 2 years of high school, known in Britain as the sixth form, students were usually appointed as "prefects," a supposed distinction that gave them authority to maintain order amongst the younger boys. My fellow Omegas and I unilaterally declined this responsibility, in conformity with our rebellion against any level of authority! One of my friends, who was skilled at restoring old automobiles, bought an enormous Mark 7 Jaguar for five pounds, in which he and I regularly skipped classes, driving out of the school car park at high speed in this huge green car. Of course, our "bunking" eventually got noticed and led to the inevitable showdown with the headmaster.

This new-found liberation notwithstanding, when the time came to apply to university, my career ambitions severely limited my options. In 1960s Britain, the procedure for college-bound students was to prioritise six university choices on an application form submitted to a centralised admissions authority. My applications listed just three, the only institutions in the United Kingdom at that time to offer astronomy as an undergraduate major. I simply left the other boxes of the application form blank. Although I was warned that this was a risky strategy, it certainly indicated I knew what I wanted to study. University College London (UCL) was the most appealing choice and ranked number one, and I was fortunate to be admitted as a first-year student in October 1968.

Life in London had many attractions, one of which was the opportunity to meet and interact with the professional astronomical community in the nation's capital. Undergraduates in UCL's astronomy programme were encouraged to attend monthly meetings of the Royal Astronomical Society (RAS), where professional astronomers gathered once a month in Burlington House, Piccadilly. However, this was "Swinging London" in the late 1960s at the height of the era of hippies, psychedelic drugs, and rock music. Whereas some university lecturers had adapted to teaching long-haired students, my reaction to seeing professional astronomers at RAS meetings for the first time was that they looked astonishingly dull and old-fashioned. Their tweed suits, ties, and, to my mind, excessively formal and humourless demeanour did nothing to enliven the atmosphere in meetings that I found to be unbearably stuffy. Did I really want to spend my life working with these people?

Inevitably, almost all participants were male, and their talks mostly revolved around the life cycle of stars or, more appealing to me, theories of cosmology—the nature and evolution of the universe on large scales. If, occasionally, someone presented results based on actual observations with a telescope, it was usually on what I considered to be a mundane topic, such as the varying light output of an individual star. Young as I was, I began to sense a disconnect in Britain's professional community between the theorists, aspiring to address the big astrophysical questions, and the observers, who, it seemed to me, were content to study minutiae.

While I pondered my future prospects, I was learning the techniques of observing. UCL has a well-equipped teaching observatory at Mill Hill, a leafy North London suburb, where we undertook observational projects once a week (plate 2). Although the London skies were as cloudy as the Welsh ones much of the time, we students could still undertake "cloudy night experiments" based on analyses of photographic plates previously taken by cameras attached to the Mill Hill telescopes. But when the weather was clear, taking and analysing my own photographs was inspirational! The largest telescope available for students was—and still is—the Radcliffe 24-inch (60 cm) refractor. I used it to photograph our Milky Way galaxy's nearest large neighbour, the Andromeda spiral (Messier 31), and to study the outermost layers of the sun during a partial solar eclipse (plate 2). Undertaking observations at Mill Hill was addictive. Although cosmology continued to have its attractions, I had no doubt that my future lay with observational astronomy.

I began my undergraduate studies at a time when astronomers were beginning to exploit wavelengths beyond the familiar optical region, which had been the sole province of telescopes back to Galileo. In 1800 William Herschel, one of Britain's most famous astronomers, discovered in a laboratory experiment that there were invisible "calorific rays" that could be reflected and refracted just like optical light. This *infrared* radiation has a wavelength longer than that of visible light and is emitted from objects cooler than the sun. Around the same time, Johann Ritter, a German chemist, conducted experiments with chemicals that reacted to sunlight and found "chemical rays" that extended to shorter wavelengths, which we now know as the *ultraviolet*. These pioneering experiments eventually led to the far-reaching discovery that celestial objects radiate across a much wider range of wavelengths than the narrow band accessible to our human eyes, from X-rays at the shortest wavelengths to radio waves at the longest.

In the 1960s, radio astronomy had emerged as a particularly active research field in the United Kingdom, as the immensely successful deployment of radar in wartime was adapted to peaceful uses, including studies of the cosmos. Radio telescopes had been built at Jodrell Bank, near Manchester, and at Cambridge University. I certainly heard a lot about their observations as an undergraduate, including the discovery of pulsars by Antony Hewish (1924–2021) and his graduate student Jocelyn Bell. Pulsars, the remnants of massive collapsed stars, are rapidly rotating compact objects that emit regular pulses of radio waves from their magnetic poles, rather like a celestial lighthouse. However, unlike optical and radio studies, making successful observations across much of the electromagnetic spectrum necessitates getting above the Earth's atmosphere since, fortunately for the human race, harmful X-rays and ultraviolet rays are absorbed by it. The same is largely true of the more benign infrared radiation. The 1960s saw the launch of modest telescopes aboard both high-altitude balloons and rockets to explore the sky at these new wavelengths for the first time. Perhaps because I was so inspired by the use of the optical facilities at Mill Hill, it never occurred to me to move into these new areas. There was enough excitement with traditional optical astronomy.

In my final undergraduate year, astronomy students had to produce a short dissertation on a research topic of their own choice, the idea being to give students a feel for what it is like to conduct original research. For my topic I chose quasar absorption lines, which offered an ingenious new means of studying the universe by analysing the nature of light emanating from its most distant objects. I was intrigued by the idea that one could study phenomena at enormous distances, well beyond the confines of our own Milky Way galaxy, in what astronomers called the "extragalactic universe."

This investigation opened up a whole new world for me. Quasars—short for quasi-stellar objects, or QSOs—were discovered in the early 1960s. Their large recessional velocities indicated they were being seen at enormous distances (more on this in chapters 2 and 3). Although their precise nature remained a mystery for many years, we now know that these exotic objects are spectacularly luminous galaxies whose nuclei harbour massive black holes. With masses often a billion times or more that of the sun, these heavyweight black holes are capable of accreting large amounts of gaseous matter from the rest of the galaxy through their dominant gravitational influence. As this gas spirals inwards, it releases copious amounts of radiation, which can be used to probe the nature of the *intergalactic medium*—the tenuous clouds of hydrogen gas and other material that fill the cosmos between galaxies.

How does this work? As the light from a distant quasar makes its way to a telescope, it intercepts clouds of intergalactic hydrogen. Although these clouds do not emit their own light and are, therefore, from the astronomer's perspective "dark," they are capable of absorbing portions of the quasar light at a particular wavelength through atomic interactions with the light particles (photons). A spectrum of the quasar reveals these absorption signals as cosmic "fingerprints" that contain valuable information about the properties of these clouds, including their chemical composition and distribution in space. Through this type of detective work, remote and otherwise inscrutable tracts of the universe become accessible to analysis. One can think of the quasar in some sense as a distant car headlight that is bright enough to reveal otherwise invisible wisps of mist rolling along the road towards you.

Atomic spectroscopy was taught at UCL by a talented and disarmingly modest assistant professor (a "lecturer," in UK academic parlance) named Bill Somerville. In his precise and soft-spoken accent (a curious blend of Scottish and Irish), he explained the mathematics of this phenomenally powerful tool of the astronomer. In an instrument called a spectrograph, through the application of a prism or diffraction grating, the light from a celestial object can be dispersed into its constituent wavelengths. This "spectrum" has a much higher fidelity than the colours visible in a rainbow and can reveal a wealth of information about the chemical and physical make-up of stars, galaxies, and the intergalactic medium. My introduction to spectroscopy's potential for probing the far reaches of the universe left an impression not unlike my earliest stargazing experiences, except here, I realised, was a tool far more formidable and sophisticated than the primitive backyard telescope that first introduced me to the night sky.

I spent nearly all my spare time poring over the latest astronomical journals in the polished wood surroundings of UCL's main library, immersing myself in this fascinating new topic. The key question for astronomers, and the topic of my project, was, exactly where was this absorption in the quasar light occurring? Was it happening in the gaseous clouds in the immediate vicinity of quasars or in the vast intergalactic spaces in between galaxies? Today we know that the answer is the latter, but that was far from clear at the time. Some astronomers even questioned whether quasars were truly energy-emitting sources at great distances; conceivably, they argued, quasars were nearby sources expelled at high velocity from our own Milky Way.

I was discovering a fascinating topic at the frontier of knowledge. None of my UCL lecturers seemed familiar with my topic of research, and yet the pace of discovery was rapid; every new issue of the *Astrophysical Journal*, a premier research publication in astrophysics, contained articles with new data. Significantly, the progress was almost entirely observational. Without exception, the quasar spectra came from telescopes in the United States, most notably from the renowned instrument where quasars were first discovered—the 200-inch Hale Telescope on Mount Palomar near San Diego, California. Although I found some theoretical papers on the topic authored by UK astronomers, they were primarily concerned with interpreting the data taken by their American counterparts This dichotomy between British and American astronomers reflected the simple fact that without access to their own large telescope, British astronomers could not lead observational campaigns at the frontier of knowledge such as those that inspired my undergraduate dissertation on quasar absorption lines.

How could this situation, so detrimental to British astronomy, have arisen? The story dates back to the late 1920s, when the giant 200-inch telescope at Palomar was conceived by the visionary solar astronomer and indefatigable fundraiser George Ellery Hale, whose role in establishing Southern California's Mount Wilson Observatory and co-founding the California Institute of Technology in the early part of the twentieth century would greatly influence the course of American science. Following its completion in 1948, the Hale Telescope (named for its originator) reigned as the world's largest and most powerful optical telescope for the next four decades. It is no exaggeration to say it dominated the field of observational astronomy (chapter 3).

In 1946, in recognition of the 300th anniversary of Isaac Newton's birth, the Royal Society of London announced plans to fund a 98-inch telescope, a facility that would have seven times the *light-gathering power* of Britain's largest telescope at the time. This term refers to an optical telescope's capacity for collecting photons. The larger the area of the telescope's primary mirror, the more photons it is capable of accumulating. The hope was that this Isaac Newton Telescope (INT) would go some way towards rectifying the depressing fact that, as far as optical astronomy was concerned, US astronomers were making nearly all the observational discoveries. Indeed, I later discovered many of those "US astronomers," including the ones pioneering the study of quasar spectroscopy, were like myself born and educated in Britain. They had emigrated to the United States when they realised there were no professional prospects for them in their native country.

It was agreed that the INT would be operated and maintained by staff at the Royal Greenwich Observatory (RGO). Astonishingly, by the late 1950s, more than 10 years after it was first envisioned, there had been little progress in constructing the telescope, other than acquiring a free mirror blank (i.e., unpolished glass) originally intended for a telescope in Michigan that was never built. Following the appointment in 1956 of a new energetic Astronomer Royal, Sir Richard van der Riet Woolley (1906–1986), construction eventually progressed. However, soon after, the bewildering and fateful decision was made to locate the new telescope alongside the RGO, which had recently been moved from London to Herstmonceux Castle in Sussex (plate 3). Apparently the proximity of RGO staff to the telescope was a more important consideration than avoiding the infamous English weather. Although some had argued that Sussex was the sunniest part of England, they failed to notice that sea mist regularly rolled in from the coast at night. These problems immediately became apparent upon the telescope's completion in 1965 and its subsequent opening in a ceremony with Queen Elizabeth II. Plans to have Her Majesty view the planet Saturn and its regal rings on that occasion had to be scrapped because it was raining. This did not augur well for the revival of British observational astronomy!

As a result of its poor location, the INT at Herstmonceux was not a great success. Indeed, I wasn't even aware of its existence until the early 1970s, by which time there were serious discussions about moving it. Ultimately, under pressure from the British astronomical community, and in a remarkable admission of failure, the INT was disassembled, transported across the sea, and then reassembled, "brick by brick," as it were, on the island of La Palma in the Canaries in 1979. In the early 1980s I was appointed to commission two new instruments on the relocated telescope (see chapter 5).

Such was the state of UK observational astronomy after I graduated from UCL and began a PhD in astrophysics at Oxford in 1971. My UCL research project had definitely fired up my enthusiasm for observations of distant extragalactic sources, but where would I get the relevant data to continue such studies? Oxford's only extragalactic astronomer, John Peach, was no longer taking graduate students for exactly this reason. Instead, for my PhD thesis work, I was redirected to a project researching the atmospheres of the sun and a bright star, Arcturus, that was within reach of the smaller telescopes to which Oxford had access. Britain was continuing to produce talented astronomers and theoretical research was progressing well, but, for the foreseeable future, those seeking observational data for world-class projects would have to emigrate to the United States. There was, seemingly, nowhere else for them to go.

2

Grand Time Machines

A night on a large telescope is a wondrous experience. Although access to the telescope comes only after much preparation, before a night's observing begins there's a brief opportunity to relax and be inspired by the panoramic view as the sun sets and darkness inexorably arrives on a remote mountaintop, often far from home (plate 4). During those precious moments, it's easy to believe that an astronomer has the best job in the world.

How does a professional astronomer secure observing time on a large telescope? Six to nine months earlier a detailed proposal is written, which involves presenting a scientific case and a technical feasibility calculation. Much thought has to be given as to the timeliness of the required observations, the importance of the expected scientific results, and how to phrase the proposal to beat competing teams who might be suggesting a similar idea. A committee drawn from the astronomical community reviews all the proposals, and there are often a hundred or more for each telescope. Telescope time is usually oversubscribed three to five times, so painful choices have to be made. Against such odds, it's inevitable that, more often than not, the proposal is rejected. The committee's decision and reasoning, or "proposal feedback," is usually sent by email. The astronomer's response to failure is predictably bitter: the committee must be composed of goons who clearly misunderstood what was a brilliant proposal. Life goes on and the astronomer waits until the next call for proposals, six or twelve months later, to try again.

If, however, the proposal is successful, the observatory draws up an observing schedule indicating the dates when the astronomer should be at the telescope. Then preparation begins in earnest, flights and accommodation are booked, and the journey begins—hopefully without snags such as cancelled flights, airline strikes, and hurricanes. After arrival at the observatory and checking in at its associated dormitory accommodation, a staff member or "support astronomer" who works at the observatory reviews the observing plans with the visiting astronomer(s). In the afternoon before the first night's observations, the visiting observers use their chosen instrument to take various calibration exposures that will enable conversion of the night's raw observations into more useful data that can provide astronomical results. There's time to grab a quick dinner (provided by an observatory canteen or, more often, self-made in a small kitchen) before awaiting darkness.

In the 1980s I was savouring that brief moment of twilight in Australia after all this preparation and the beginning of observing. The Anglo-Australian Telescope has a catwalk; this is an elevated platform that encircles the dome from which the approaching weather can be judged and, of course, from which a sunset can be admired. The observatory also has a visitor centre, and tourists often stay after it closes to witness the dark night sky. On this occasion, one tourist shouted up to me, "How did you get up there?" implying by what means of elevators or steps did I get where I was. I had to reply, "I studied hard for a decade!"

Since the 1970s, large telescopes and their scientific instruments—usually a camera or a spectrograph—have been controlled by software. The control of the telescope is usually the responsibility of a resident night assistant or telescope operator. He or she opens the telescope dome and also has the ultimate authority for closing it if the weather conditions deteriorate. The operator issues the commands to move the telescope and locates a nearby "guide star" when the astronomer's target is located and ready to be observed. Accurate guiding is essential for long exposures taken with a scientific instrument because a telescope's own tracking is never good enough. The light signal received from this guide star is continuously monitored and is used to make small adjustments to ensure the telescope precisely tracks the target. The astronomer, on the other hand, is usually in control of the scientific instrument, deciding the sequence of targets and the duration of each exposure and judging whether adequate data are being taken. Should technical issues arise, the support astronomer is on call. A few brave astronomers observe single-handedly, but most arrive in teams of two or three so that one of them can operate the science instrument while the others scrutinise the incoming data. Although observatory control rooms have changed their user facilities over the years owing to advances in computing (see plate 5), the overall arrangement remains the same. Observers sit alongside the night assistant or communicate via a remote video connection, and, if things go well, everyone finds time to establish a friendly rapport.

Over the years I have clocked up nearly 800 such nights, amounting to around 5% of my working life, spent thousands of miles away from home on isolated mountaintops with small bands of like-minded skywatchers. Under ideal circumstances, there is a real sense of camaraderie in the control room as the telescope operator, the support astronomer, and the observing team work harmoniously and efficiently together to gather unique and sometimes unexpected data. In my experience, a graduate student can learn an enormous amount from being at the telescope. In addition to enjoying the undivided attention of their PhD supervisor and witnessing at first-hand how data are acquired and evaluated in real time, the experience is inspirational.

The excitement of a successful night on a large telescope is uplifting and addictive. The ultimate "high" is a discovery made in real time— perhaps finding a surprising or new class of celestial source, or realising from a preliminary assessment of the data that, in all likelihood, the question posed by the original proposal is about to be dramatically answered. This is a cause for team celebration, phone calls to colleagues at home, maybe even champagne at dawn, as the exhausted team retires to the dormitory buoyed by its probable success. Sadly, however, the opposite is just as likely: a cloudy night after travelling, in some cases, halfway around the world or, even worse than overcast skies, strong winds, high humidity, or a technical issue that prevents observing on an otherwise beautiful, clear night (plate 6, *left*). It's very frustrating, and the poor graduate student whose PhD thesis depends on securing all-important observational data often becomes despondent. The team stares gloomily at an image from an all-sky camera that, despite its state-of-the-art design and latest accoutrements, shows only fog or cloud. Most observatories are unable to offer compensatory nights for an unsuccessful observing run for the simple reason that every single night of the year has already been allocated to another group of astronomers. The team often flies home with no results, facing the prospect of starting all over again with an updated proposal and the wait for committee approval. Observational astronomy can be a rollercoaster of mixed emotions, but it is certainly character-building!

Over my career, there have been significant changes in how observing is conducted, and, sadly in my view, the hands-on mode of observing described above (sometimes called "visitor mode") is becoming rarer. Improved computational facilities and internet speeds mean that astronomers can control their scientific instruments from afar—either from a dedicated facility in their own institution or even, as I did recently during the Covid-19 lockdown, from a laptop at home. The romantic concept of the intrepid astronomer preparing to make inspiring, even breathtaking, discoveries at a dramatic location beneath a star-studded night sky is rapidly being replaced by yet another day in the office with a computer screen and a video link. There is also the sad fact that this trend for "remote observing" is depriving observatory personnel of the opportunity to meet and work with visiting astronomers, for which generous library facilities were traditionally provided (plate 6, *right*). They can't easily share the observers' enthusiasm over discoveries made possible through the staff's dedicated on-site work at the observatory. Camaraderie and excitement are dampened and the entire experience is less personal. Nevertheless, the trend towards remote observing seems inevitable given the high cost, inconvenience, and climatic impact of long-distance air travel. It does also make observing more inclusive for astronomers who may not have the funds to travel to the observatory. Although these are legitimate trade-offs, I can't help thinking that something invaluable is being lost.

Let me now turn to the progress in telescopes and their performance over the past century and why the largest telescopes have featured so prominently in my career and those of many of my colleagues. The performance of a large telescope is governed by several factors. The most important factor is the size of the main, or primary, mirror. Refracting telescopes, which only use lenses, such as the Radcliffe Telescope at Mill Hill in London to which I had access as an undergraduate (chapter 1), dropped out of favour a century ago because glass lenses whose diameters are greater than 1 metre are hard to manufacture and difficult to support when the telescope is moved from one sky position to another. From the 1920s onward, reflecting telescopes with mirrors have become standard. These mirrors begin as glass blanks, cast in a furnace and allowed to cool slowly. There follows a sustained period of painstaking polishing to create a parabolic or hyperbolic mirror surface, whose form is necessary to bring the light of distant objects into focus. Once the shape has been tested and found to be satisfactory, the concave surface receives a thin coating of reflective material such as aluminium or silver. This large mirror is now ready to gather the light from celestial objects and bring it to focus at the entrance aperture of an instrument that is usually mounted on the telescope.

The light-gathering power, introduced in chapter 1, of any primary mirror increases with its surface area and hence the square of its diameter. For most applications, this means that a telescope whose mirror is 8 metres in diameter has four times the light-gathering power of one with a mirror only 4 metres across. Consequently, this means that a research project that would take one night on an 8-metre telescope might take four or more nights on a smaller one, assuming of course the weather remained stable through the observing run. In practice, the most ambitious projects can usually be accomplished only on the largest telescopes.

Larger mirrors are harder to cast and polish than smaller ones, and they also require bigger and more expensive laboratory facilities to house them. They are also more difficult to transport to remote mountaintops. Nor do these challenges end once the mirror is installed in the telescope. A sizeable single, or "monolithic," mirror distorts under its

own weight as the telescope tracks the sky, and correcting for this flexure requires an elaborate support system within the telescope. These
difficulties led the scientists and engineers who worked successfully on
the design and engineering of a new generation of 8-metre telescopes
in the 1990s to conclude they had reached the maximum practical size
for a single mirror. Segmented mirrors were the answer. The twin Keck
10-metre telescopes in Hawaii had already pioneered the concept of a
segmented primary mirror in the late 1980s. Both Keck mirrors com-
prise 36 individual hexagonal mirrors, each segment supported independently in order to orchestrate a single reflecting surface (plate 7; see
also chapter 7). For the even larger telescopes under construction today,
the segmented-mirror approach is now widely viewed as the only way
forward.

In studying a faint star or galaxy with a large telescope, a second
factor governs the performance. The night sky is not completely dark,
as the tenuous upper layers of the Earth's atmosphere radiate. This radiation arises as a result of both the temperature of the atmosphere and the
"airglow" from its atoms and molecules, which are excited by solar radiation and various extraterrestrial particles. For an amateur observer looking at very bright stars through binoculars or a small telescope, this
uniform "sky background" is barely noticeable and does not detract
from the observations. However, for a professional astronomer studying
exceedingly faint galaxies, the sky signal entering the instrument may
be several hundred times brighter than that of the distant celestial object
under investigation. Removing this sky background is therefore crucial
to progress. Even a small error in cancelling it out can lead to a residual
signal that swamps that of the target. It's like trying to hear the music
played by one violin in the cacophony of a full orchestra performing
Tchaikovsky's 1812 Overture.

Various methods exist for removing the unwanted sky signal, the sim-
plest being to move the telescope back and forth from the target to a
blank patch of sky and subtracting the difference in light signals entering
the instrument. This "nodding" of the telescope is particularly necessary
for astronomical spectroscopy, where the light of a galaxy is examined
over a range of frequencies or wavelengths. In this case, the airglow from

the excited atoms in the upper atmosphere is especially troublesome because it occurs at specific frequencies, or wavelengths, and its strength can vary with time and sky position.

The final factor governing the performance of a telescope is the efficiency of the scientific instrument attached to it. Astronomers use cameras to take pictures of the night sky, locate their desired objects, and measure their positions. We use spectrographs to break the light of such objects into spectra that can be used to determine their chemical composition and other properties in greater detail. Over the last 40 years, there has taken place a remarkable technical revolution in the performance of these astronomical instruments. In addition to improved optical components, the greatest advances have come in the design and engineering of digital detectors. As recently as the 1970s, the detector used in both cameras and spectrographs was the photographic plate, a glass plate coated with a light-sensitive emulsion (plate 8, *left*). In the same way as all photographers in the past worked, the photographic plate was loaded into a wooden or metal holder and mounted at the focus of the camera or spectrograph affixed to the telescope. A metal plate or shutter was then withdrawn to begin the exposure. After the normal processes of developing and fixing the emulsion, the photographic plate was scanned with a "densitometer" that used a photocell to measure the amount of blackening on the negative image. And the scanned output was then stored on paper. The efficiency of such photography, measured by the fraction of incoming light particles, or photons, that registered with an adequate signal on the final paper tracing, was only between 1% and 3%. In the first part of the twentieth century, exposures over several nights were necessary to study the most distant galaxies known at the time. The observer commenced taking the exposure at the beginning of the night and closed the shutter at dawn. The process resumed the following night, after setting the telescope to exactly the same position.

Milton Humason, observing assistant to the famous Edwin Hubble (1889–1953), whose name graces the Space Telescope, once undertook a multinight photographic exposure on the 100-inch Hooker Telescope on Mount Wilson in California, but, unfortunately, when it came to developing that precious photograph he inadvertently placed it in the fixer bath before the developer bath. A gentleman through and through, instead of shouting a profanity when he realised his error, it is claimed he simply said, "Oh dear."

In the early 1980s a new type of digital detector, the charge-coupled device (CCD), arrived at many telescopes. First developed at Bell Laboratories in the late 1960s, the CCD detector is an array of metal-oxide semiconductors, each pixel of which acts as a capacitor for storing electric charge. Incoming photons induce an accumulation of electric charge in the pixels, and an external circuit reads the integrated charges into a computer at the end of the exposure. Early CCDs were limited in format to around 500 by 500 pixels and had efficiencies of 30%–50%. Again, these efficiencies relate to the fraction of incoming photons that register a measurable signal on the detector. Immediately, telescopes equipped with CCDs realised enormous gains over those with photographic plates, equivalent in some applications to having a telescope with a mirror five times larger. CCDs with larger formats, and hence many more pixels, can now be mosaicked to make panoramic cameras with huge fields of view and detector efficiencies approaching 90% (plate 8, *centre*).

I still vividly recall my first observing experience with a CCD. This was in the early 1980s, and I was on an observing run at the Anglo-Australian Telescope, where the renowned photographer David Malin happened to be present. Malin, originally a chemist, was famed for the beauty of his astronomical images and had a particular skill for enhancing his photographs to reveal faint features that would otherwise be invisible. This was also his first experience of witnessing CCD imaging. As my data came in, he looked sceptically at the computer screen displaying the digital image I had just taken of a cluster of galaxies. I supposed that he recognised there and then that this detector represented the future of astronomical imaging and he was going to be out of work. To my surprise, however, he started taking photographs of the screen with his film-based camera, commenting that he believed he could further enhance these digital images!

Technology, therefore, has governed the development of observational astronomy since I began my professional career in early 1970s. In the upcoming chapters I will trace this story in the context of a major theme: directly studying the birth and evolution of galaxies using large, powerful telescopes as "time machines" to observe and study the universe as it looked in its distant, even primordial, past. The ultimate goal is to look back to see the birth of the very first galaxies—the so-called *cosmic dawn.*

The concept of witnessing the past will be familiar to anyone who has looked at different geological layers of rock on a cliff, or counted growth rings on the felled trunk of an ancient tree, or even studied an old family photograph. For the astronomer, the crucial aspect in this regard is light, which travels with a finite speed of nearly 300,000 kilometres per second. What this means is that when we view an object separated from us by any distance, we are seeing it as it appeared in the past, whether that past is measured in a fraction of a microsecond or in billions of years. The sun is just over 8 light minutes away, and Earth's nearest star, Proxima Centauri, is 4 light years distant; so we see the sun and Proxima Centauri as they were 8 minutes and 4 years ago, respectively. Starlight emanating from the centre of our Milky Way galaxy takes about 27,000 years to reach us, and light from our nearest large neighbouring galaxy, the Andromeda spiral (pictured in plate 8, *right*), takes nearly 2.5 million years. But even these "lookback times" are modest on cosmic scales. Astronomers do not expect nor detect significant evolution in the universe over such short timescales. The most recent reliable estimates put the age of the universe at approximately 13.8 billion years, so if this was represented by a human lifespan of 80 years, we'd be witnessing the Andromeda spiral as it was only 5 days ago. From our understanding of how stars form, evolve, burn their nuclear fuel, and die, the evolution of galaxies made up of billions of stars is expected to proceed on timescales of a billion years or longer.

Astonishingly, our modern large telescopes are now sufficiently powerful to take images that reveal galaxies at far greater distances, corresponding to lookback times of many billions of years. We thus have the ability to identify and study galaxies that existed when the universe was only a small fraction of its present age. Such an image (plate 9) can be considered a "time tunnel," a way to see galaxies over a range of distances and therefore different lookback times. This raises two important points. Firstly, we cannot observe evolution in an individual galaxy. Our own lifespan is so insignificant compared with a galaxy's likely evolutionary timescale of at least a billion years that we are able to decipher evolution only by studying the population as a whole. The various galaxies in the image are each seen at their own, different, phase in cosmic history. We somehow need to connect their properties into a single story, much as a visiting alien might have to deduce the life cycle of a human by witnessing, in a single moment, images of children, adults, and senior citizens. Secondly, to infer this evolution of the galaxy population, we clearly need some way to identify at what period in cosmic history each of the galaxies in our picture is being observed. In other words, we need some way to accurately estimate the distance to each galaxy so we can determine how far back in cosmic time we are witnessing it.

Let's now turn to how astronomers estimate distances and, for galaxies, their lookback times. Measuring astronomical distances is always a challenge. For nearby stars in the Milky Way we can use triangulation, exploiting the different perspective the Earth provides as it moves in its orbit around the sun. Compared with a backdrop of distant stars, a nearby star will apparently move by a small amount on the sky, reflecting the Earth's motion during the year. As long as we know the radius of the Earth's orbit about the sun, the apparent angular motion of the nearby star on the sky enables us to estimate its distance. Beyond such nearby stars, however, much of twentieth-century astronomy has been concerned with trying various other methods to arrive at accurate absolute distances, usually calibrating a method applied to distant sources with one used for closer sources. Fortunately, for the galaxies pictured in our "time tunnel," the lookback times, or cosmological distances, are so great that the universe itself has evolved during the light travel time. We shall see below that it has expanded in size and we can exploit this fact to provide the time at which the light left a distant galaxy.

The expansion of the universe was discovered in the late 1920s, when it was realised that most nearby galaxies are moving away from us. These velocities were measured largely by Vesto Slipher (1875–1969), who used a spectrograph at the Lowell Observatory in Flagstaff, Arizona to detect the galaxies' motions using the so-called "redshift" to longer wavelengths of the atomic lines of hydrogen and oxygen. By 1917 he had found that out of 25 nearby galaxies, 21 were moving away from us. Remarkably, at this time, many distinguished astronomers were still not convinced these galaxies, then referred to as "nebulae," were physically beyond the Milky Way. By 1929, many of the distances to Slipher's sample had been measured by Edwin Hubble. To estimate his distances, Hubble located variable stars in these galaxies and measured the period of variability for each one. Such "Cepheid" variable stars are known in the Milky Way and display a correlation between the period of variability and their intrinsic luminosity. This class of star is actually pulsating, with its radius increasing and decreasing with time in a regular manner. We now understand why the rate of pulsation depends on luminosity. Just as it is possible to estimate the distance to a light if the nature of the emitting source is known, so by comparing the luminosity of a variable star inferred from its period of pulsation with how bright it appears at the telescope, its distance can be determined.

Hubble claimed that there was a clear trend between the velocity of recession, or redshift, determined by Slipher's observations and the distance he deduced from his Cepheid variable stars. The trend was in the sense that the more distant galaxies are moving away from us faster, and this was later interpreted as evidence for an expanding universe. This correlation between increased velocity and larger distance became known as Hubble's law, and the slope, or gradient, of the relation, which effectively measures the rate of expansion today, is known as Hubble's constant.

In recent years, historians of science have begun to question whether Hubble in fact deserves sole credit for discovering the expansion of the universe. They point out that his 1929 scientific article does not actually claim that the universe is expanding; he was surprisingly cautious in interpreting his velocity-distance relation. Moreover, we now know that his distance measurements were seriously flawed, both on an absolute scale owing to an incorrectly applied local calibration for the Cepheid variable stars and because his largest distances were underestimated.

Even the trend that he claimed to have discovered was of marginal significance. A contemporary reanalysis has suggested that a sample of galaxies at twice the distance limits of Hubble's sample would be needed to robustly detect the cosmic expansion. Finally, Hubble in his 1929 paper gave Slipher little credit for providing the all-important recessional velocities. A modest astronomer, who published in internal reports from his observatory, Slipher outlived Hubble by 16 years but has only recently received the recognition he deserves for his pioneering contributions to what eventually became accepted as our modern view of the expanding universe.[1]

It is incorrect to picture the expansion of the universe as galaxies moving as projectiles through existing space. According to Einstein's general theory of relativity published in 1917, which describes the behaviour of space in terms of the density of the material it contains, *it is space itself that is expanding.* The application of Einstein's theory to the universe at large readily suggests the possibility of such an expansion, but Einstein was not ready for such an inference and fudged a solution to maintain a static universe. To oppose the effects of gravity, he introduced a repulsion term into his equations, for which there was not, at the time, any physical evidence. He later greatly regretted this step and repudiated it. Other early cosmologists, including Hermann Weyl (1923), Ludwik Silberstein (1924), and most notably a Belgian priest, Georges Lemaître (1927), deduced that the expansion of the cosmos was a natural outcome of Einstein's theory, and they even predicted the velocity-distance trend eventually claimed by Hubble. Lemaître's article contained the most well-developed arguments, but it was published in French in an obscure Belgian journal, and in his 1929 article Hubble did not reference it. In 2012, the International Astronomical Union, the global association of professional astronomers, voted to rename Hubble's law the Hubble-Lemaître law.[2]

1. I attended a conference held at Flagstaff, Arizona, in September 2012 to celebrate 100 years since Slipher's first measurement of the velocity of a spiral galaxy.

2. The International Astronomical Union held a members' ballot on renaming Hubble's law, and 20%, including me, voted against the motion. The majority of members no doubt accepted the need to give appropriate credit to both scientists. In my case, my vote was not due to

Returning to my goal of studying galaxy evolution, how does the expansion of space provide a practical means of estimating the distance and the time at which we are observing a remote galaxy? Slipher's galaxies had recession velocities of only a few hundred kilometres per second so they were not much further away than the Andromeda spiral. He deduced their motion by the shift of characteristic atomic spectrum lines to redder wavelengths seen in their spectra. As we probe to greater distances with today's more powerful facilities, the recessional velocities, or redshifts, become much larger. As galaxy distances soon run into trillions of kilometres or millions of light years, astronomers prefer to use the label of "redshift," commonly denoted by the letter z, as a distance indicator. We will use this letter z for redshift throughout this book. This redshift is simply a measure of the amount by which a spectrum line has been shifted, expressed as the ratio of the received wavelength to the wavelength originally omitted, as determined in the laboratory. Here are two examples. A redshift $z = 0.5$ implies that the wavelength of a familiar spectrum line—for example, due to oxygen or hydrogen—has been shifted longward to a new value $1 + z = 1.5$ times its laboratory value. A redshift of $z = 1$ would imply a shift to a wavelength twice $(1 + z = 2)$ its laboratory value.

However, this redshift represents much more than a convenient accounting tool for astronomers. Recall that the galaxy velocities should not be interpreted as akin to projectiles moving through space. It is the *stretching of space* due to the cosmic expansion, and its effect on light rays, that causes the redshift. Consider a blue light ray leaving a distant galaxy towards the Earth (plate 10, *top*). During its long journey, this light ray is stretched to a longer (redder) wavelength by the factor that the universe has expanded in the meantime. This means that the redshift (z) is something quite profound. It is a measure of the factor $(1 + z)$ by which the universe has grown in size since the light left the galaxy. If we

---

disrespect for Lemaître nor indignation that Hubble was being marginalised, but a reluctance to begin retrospectively correcting attributions, a process that could gather momentum and lead to the widespread renaming of scientific laws and constants, and even observatories and institutions.

could construct a model of how the cosmic expansion proceeds with time, the redshift would give us a direct measure of the lookback time to any galaxy (plate 10, *bottom*).

You might reasonably ask whether we understand the past history of our universe sufficiently well to convert redshift into lookback time. In the early portion of my career, this was indeed a problem. Until the early 2000s, cosmologists argued a lot about the history of the expansion of the universe. Once Einstein accepted in the early 1930s that the universe was expanding, his general theory of relativity predicted that the rate of expansion would be slowed down by the gravitational attraction of matter within it. But the amount of cosmic deceleration, as this slowing down became known, would depend on exactly how much gravitating material there is. Attempts to measure the density of matter in the universe resulted in the discovery that much of the material is dark, and likely not the normal material that makes up you and me and our familiar surroundings. The physical nature of what is now known as "dark matter" remains a mystery. An even bigger surprise came in the late 1990s, when two teams of astronomers, one of which included myself, both discovered that the expansion is currently *accelerating* (discussed in more detail in chapter 3). The teams compared the rate of expansion today with that in the past, effectively extending the Hubble-Lemaître law to earlier times by using supernovae—exploding stars—as distance indicators. This cosmic acceleration could not be explained by the attractive force of gravity and so its discovery indicated a further new ingredient of the universe—given the moniker "dark energy"—which is, perhaps, a property of empty space that provides a natural repulsion.

Given two major mysteries facing cosmologists—dark matter and dark energy—it may seem a hopeless task to predict the expansion history of the universe and link a galaxy's redshift with the time in the past when we are observing it. However, great progress has been made in determining the quantitative contributions of dark matter and dark energy to the energy budget of the universe (even if the physical nature of both remains unclear), and this is sufficient for an accurate model of the expansion history. So, armed with a large sample of galaxies of known redshift derived from spectroscopy, astronomers today are using the largest telescopes to trace the evolution of galaxies back to when they formed.

Since a galaxy is composed of stars and gas, the latter representing the fuel from which new stars are born, we can sketch a simple picture of how galaxies form and evolve. When the universe was very young, space was dark and contained only tenuous clouds of hydrogen gas—the atoms of this most abundant element being formed of a positive proton and a negative electron. Under gravity, aided by the presence of dark matter, these clouds eventually collapsed and became sufficiently hot in their centres that nuclear fusion occurred, giving rise to stars that began to shine. It is thought that rather than forming one supermassive star, these clouds fragmented to form clusters of young stars. In some sense these star clusters can be regarded as "primordial galaxies," born at the time referred to as cosmic dawn. In contrast to our own galaxy, the Milky Way, these young galaxies were forming stars more rapidly, were more luminous and compact, and were far less massive.

As time progresses, stars themselves evolve and die, but so long as there is a continued supply of gas, new generations of stars can form. When stars above a certain mass die, they explode as supernovae—dramatic events that expel the chemical products that have been synthesised through nuclear fusion in the star during its lifetime. Hydrogen nuclei are fused to form ones of helium, helium to carbon, and carbon and oxygen to heavier nuclei such as those of iron and nickel. These nuclear products are thus mixed into the gaseous fuel that forms future generations of stars. Thus, as galaxies develop and age, their gas clouds become increasingly enriched with heavier elements. A tiny fraction of the heavier elements so formed in successive generations of stars finds its way into the calcium in our bones and the iron in our blood.

This simple picture treats galaxies as isolated systems. However, via gravitational interactions, smaller galaxies may be drawn close to, and ultimately merge with, larger ones. Also, the gas supply in a galaxy might be augmented by hydrogen gas drawn from the intergalactic medium, or lost through ejected material propelled by supernova explosions. These additional influences suggest that the immediate environment around a galaxy plays a role in its assembly history. There are further puzzles. For example, why do some galaxies have spiral arms and rotate rapidly while others are featureless and do not rotate? To address these details, the overall physical processes can, in principle, be formulated, and theoretical predictions can be made. The challenge is to verify the picture deduced theoretically with observations.

When I discuss my research in astronomical seminars, I often joke that it's an unfortunate fact that we live in a world with theoreticians. They have no need for observations and are content to run computer models to address major questions about our universe. I once told a theorist colleague of mine that I would be out of town next week as I was going observing. He asked what questions my observations were going to address. When I told him, he responded briskly that my trip was a waste of time as he already knew the answers! Contrary to what I suspect would be the preference of theoreticians, observational astronomers do not generally head for their telescopes primed with checklists of the latest theoretical predictions. Observers such as myself genuinely believe that progress is best made empirically, through carefully planned observations that also permit discoveries beyond the predictions of current theory. Indeed, the history of the subject of galaxy evolution has been driven largely by observations and technological advances, as this book will reveal. In many cases, observations have shaped our current understanding and led to the need for new theoretical models. In the coming chapters I will take the reader through the development of this story over the 50 years of my career and a relentless pursuit of the birth of galaxies.

PLATE 1. How I discovered astronomy: the little blue book (here with illustrated cover) and the inscription from its author, Sir Patrick Moore, seen with me in 2007 at a party celebrating 50 continuous years of his famous monthly BBC TV programme *The Sky at Night*.

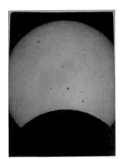

PLATE 2. (*Left*) University College London Observatory at Mill Hill. (*Centre*) The 24-inch Radcliffe refractor. (*Right*) Photograph taken by the author with the Radcliffe refractor of the February 1971 partial solar eclipse.

PLATE 3. (*Left*) Herstmonceux Castle, the home of the former Royal Greenwich Observatory and the location of the Isaac Newton Telescope in the late 1960s. (*Right*) The relocated telescope on the island of La Palma.

PLATE 4. Evening twilight on the Canarian island of La Palma with a view of the Roque de los Muchachos Observatory from the 2400-metre elevation summit.

PLATE 5. The evolving nature of telescope control rooms. (*Left*) Night assistant's console at the 3.9-metre Anglo-Australian Telescope, one of the first computer-controlled facilities. The observers sit alongside the night assistant. (*Right*) Observer's station at the Waimea, Hawaii, headquarters of the W. M. Keck Observatory, detached from the telescopes on the summit of Maunakea. In this case the night assistant at the telescope is reachable via a video link.

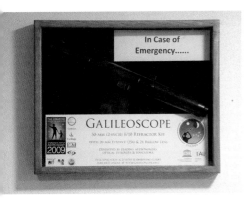 

PLATE 6. (*Left*) What to do in the case of telescope failures (control room at the Hale Telescope on Mount Palomar). (*Right*) Evidence of a bygone era—well-equipped libraries for visiting astronomers are a familiar sight at older observatories but are now rarely seen. This one is at the Las Campanas Observatory in Chile.

PLATE 7. Large primary mirrors. (*Left*) The 8-metre-diameter "monolithic" mirror of the Gemini North telescope atop Maunakea, Hawaii. (*Centre*) Keck I's 10-metre mirror is made up of 36 hexagonal segments, each separated from its neighbour by a modest 3mm gap. (*Right*) An individual Keck segment after routine recoating. It is 1.8 metres across its widest point, 7.5 cm thick and weighs half a ton.

PLATE 8. The detector revolution. (*Left*) A photographic plate taken of the Andromeda spiral galaxy in 1965. (*Centre*) The Hyper Suprime-Cam instrument on the 8.2-metre Subaru Telescope has a detector comprising a mosaic of 104 CCDs, each with a format 2048 by 4096 pixels—a total of 870 million pixels. (*Right*) Image of the Andromeda spiral taken with the Hyper Suprime-Cam.

PLATE 9. A deep exposure of a small patch of sky (less than a tenth the diameter of the full moon) taken with the Hubble Space Telescope and revealing many hundreds of galaxies over a wide range in distance and hence "lookback time." To piece together cosmic history, astronomers seek to estimate the distance and hence place a "time stamp" on each one.

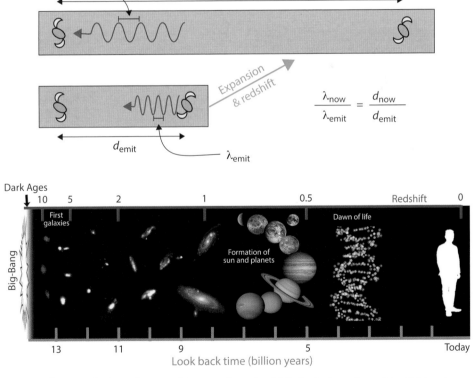

PLATE 10. (*Top panel*) Physical origin of the cosmological redshift. In the lower box, a blue light ray leaves a distant galaxy. By the time it reaches us on the left in the upper box, space has expanded and the light ray is stretched to a longer (redder) wavelength by the factor by which the universe has expanded. The ratio of the received and emitted wavelengths $\lambda_{now}/\lambda_{emit}$ is simply the factor d_{now}/d_{emit} by which the universe has expanded since the light was emitted; both ratios are equal to $1 + z$, where z is the redshift. (*Lower panel*) Schematic of how redshift relates to the lookback time to a distant galaxy.

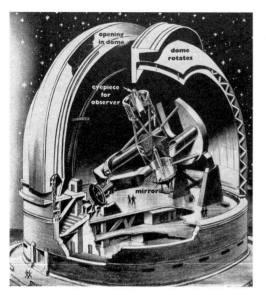

PLATE 11. (*Left*) Inspiration for an 8-year-old: an unnamed large telescope in a children's encyclopaedia. (*Right*) The Hale 5-metre (200-inch) reflector completed in 1948.

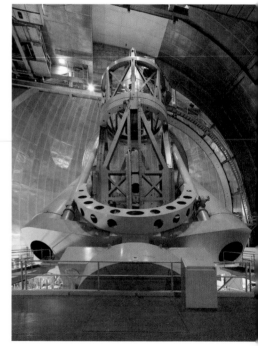

PLATE 12. (*Left*) With Allan Sandage during his visit to the inaugural UK National Astronomical Meeting at Durham University in 1992. (*Right*) By inheriting Hubble's programme in 1953, Sandage had unrivalled access to the Hale 200-inch soon after it was completed.

PLATE 13. (*Left*) With Jim Gunn in Tokyo in 2010. Gunn combined theoretical insight with observational and technical wizardry to change the motivation for studying distant galaxies. During the 1980s we were competitors in the quest for cosmic evolution but are now both involved in completing a major spectrograph for the Japanese Subaru Telescope. (*Right*) Hubble Space Telescope image of 0024 + 1654 at a redshift $z = 0.40$, one of the most distant clusters targeted by Gunn and Oke in their Palomar campaign.

PLATE 14. Twilight on Mount Palomar with the Hale Telescope ready for action.

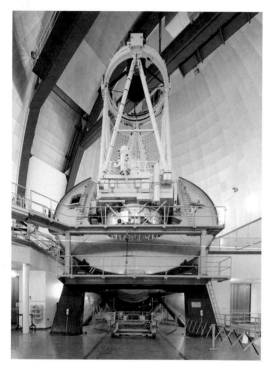

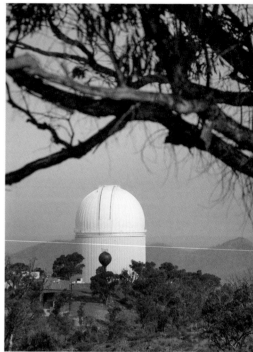

PLATE 15. The 3.9-metre Anglo-Australian Telescope (AAT). Situated in the beautiful Warrumbungle National Park in New South Wales, the AAT brought UK and Australian astronomers into the vanguard of optical astronomy.

PLATE 16. The 1.2-metre UK Schmidt Telescope on Siding Spring. Installed by the Royal Observatory, Edinburgh, it undertook a photographic survey of the southern sky and initiated statistical studies of galaxy evolution and cosmology by UK and Australian astronomers.

PLATE 17. (*Left*) The fledgling "Durham group" in 1978: from left to right, me, Tom Shanks, and George Efstathiou. (*Right*) The Automated Plate Measuring machine at Cambridge, one of two UK measuring machines built to scan UK Schmidt photographic plates.

PLATE 18. Observing in the 1980s. (*Left*) The AAT was one of the first large telescopes to be computer controlled. The night assistant operates the telescope and maintains an observing log. (*Right*) The visiting astronomer (a somewhat animated version of the author here) manages the scientific instrument. In this case the spectrograph is being used with Boksenberg's IPCS detector.

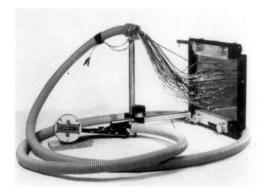

PLATE 19. Fibre-optic multi-object spectroscopy at the AAT in the early 1980s. (*Left*) Fibres plugged into a brass plate with holes precisely drilled at the positions of many galaxies are collected and aligned in a row matched to the input slit of a spectrograph. (*Right*) Peter Gray is inserting fibres into a brass plate while I note which fibre number is assigned to which hole number.

PLATE 20. Faint galaxy redshift survey teams at Coonabarabran airport. (*Left*) A playful Tom Broadhurst after a successful campaign with the multi-fibre spectrograph. (*Right*) The LDSS team, including Keith Taylor (left) and Richard Hook (right).

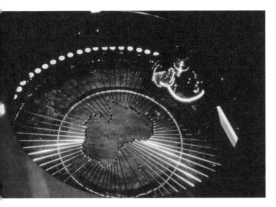

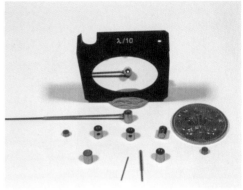

PLATE 21. The Autofib concept developed by Ian Parry. (*Left*) As a public-relations exercise, 64 fibres were arranged to form a map of Australia. A frequent annoying response was, "What about Tasmania?" The "pick-and-place" electromagnet can be seen top right. (*Right*) Autofib components. Each fibre remains on the focal plane via a tiny button containing a powerful solid state magnet, and a right-angle prism reflects incoming light into a fibre passing through a long tube. In his conference talks, Parry would joke that the British two-pence piece represented the funds remaining in the budget after construction was completed.

PLATE 22. The LDSS project—a multi-slit faint object spectrograph that was pioneering in its approach, with the largest field of view at the time. (*Left*) An aluminium mask was designed for each field, with slitlets cut for many faint galaxies and circular holes for bright stars used to accurately acquire the field. (*Right*) As it was a manual, no-frills instrument, observers had to ride in the daunting Cassegrain cage for hours at a time in complete darkness.

PLATE 23. The monster two-degree field instrument at the prime focus of the AAT. (*Left*) A "pick-and-place" robot on an *x-y* carriage can position 400 fibres, while a separate set of the same beneath is being observed. After the exposure is completed, the instrument tumbles 180 degrees and the newly configured fibres are ready to observe the next field. (*Right*) A zoom in on the fibres and their buttons. Following the original idea of Ian Parry, fibres can cross one another, improving their assignment to a complex distribution of astronomical targets in the focal plane.

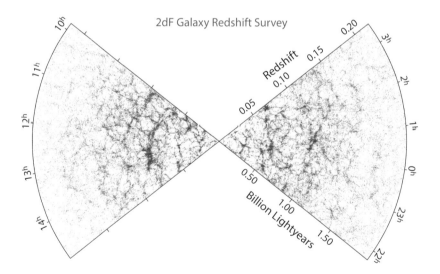

PLATE 24. The spatial distribution of 250,000 galaxies from the 2dF Galaxy Redshift Survey completed in 2003. The survey charted the galaxy distribution over unprecedentedly large scales and provided fundamental cosmological results. Each faint dot represents a galaxy whose redshift was determined using the 2dF instrument. The survey was conducted in both the northern (left) and southern (right) Galactic hemispheres. The observer is at the centre of the figure and the distance outwards in all directions represents the redshift, or distance.

PLATE 25. An evocative view of the 4.2-metre William Herschel Telescope seen against the Atlantic Ocean far below after sunset.

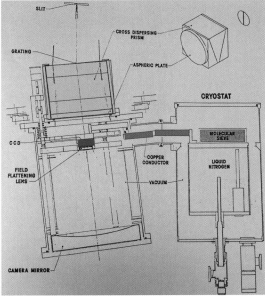

PLATE 26. The innovative Faint Object Spectrograph at the Cassegrain focus of the 4.2-metre William Herschel Telescope (*left*) and its optical design (*right*). The entire instrument is the size of a suitcase and extraordinarily efficient. The traditional collimating mirror is eliminated via correcting optics attached to the diffraction grating (green). The CCD detector (red) is cooled by a conducting "finger" (blue) at the internal focus of the camera mirror. The logo at the base of the instrument shows a Hawaiian racing tortoise with the quote, "I may look slow, but watch me go!"

PLATE 27. After nearly a decade of use at the William Herschel Telescope on La Palma, LDSS-2 was purchased from Durham University by the Carnegie Observatories, refurbished, and installed on the 6.5-metre Clay Telescope at the Las Campanas Observatory in Chile in 2001. I stand next to the relocated instrument alongside Paul Atherton, who helped design LDSS-1 at the Anglo-Australian Telescope.

PLATE 28. (*Left*) Simon Lilly, characteristically well dressed, with David Koo at a conference in Durham in 1988. (*Right*) The author with Len Cowie (centre) and Garth Illingworth at a conference in the Netherlands in the early 2000s.

PLATE 29. An early colour image of a small patch of sky taken with the Hubble Space Telescope, revealing the dominant population of faint blue galaxies.

PLATE 30. (*Left*) "Spoilt for Choice"—a cartoon accompanying an article in the *Economist* (July 14, 1990) discussing the merits of the Spanish versus the American option for UK astronomers. (*Right*) Ground-breaking for the Gemini-South telescope on Cerro Pachon, Chile, with the UK astronomers who successfully promoted the American option (from left to right: Roger Davies, Malcolm Longair, Andy Lawrence, the author, and Matt Mountain). Mountain later became, in turn, director of Gemini, director of the Space Telescope Science Institute, and the president of AURA.

PLATE 31. The northern Gemini 8-metre telescope ato Maunakea, named after Fred Gillett, the original projec scientist for the international twin-telescope project. Th United Kingdom became a founding partner with Canad and the United States following the deliberations of th LTP. The telescope design is unusual in offering only on (Cassegrain) focus.

3

Palomar

The Perfect Machine

In my family home in Colwyn Bay, a seaside resort on the North Wales coast, my older sister and I shared the *Odham's Encyclopaedia for Children* (1954). In the late 1950s, there was no internet, of course, and even black and white television was a rarity. I therefore spent all my spare time reading this lavishly illustrated book. Some years ago I managed to find a copy in a second-hand book shop for 35 pence, and even now, 60 years later, I can recognise every single page; such is the impact it had on me. The astronomy section contained an illustration entitled "Inside an observatory, with its huge telescope pointing to the sky." It had a cutaway drawing showing the large dome, the telescope and primary mirror inside the enclosure, and a place labelled "eyepiece for observer" (plate 11). I was in awe of this drawing, turning back to the page countless times. I had no idea that this was the venerable 200-inch telescope on Mount Palomar, let alone that over 40 years later I would become the director of the Palomar Observatory.

In 1928, George Ellery Hale (1868–1938) was making the case for private funds for what would become the largest telescope in the world for 45 years. A phrase he wrote in an article at the time encapsulates the theme of this book: "Like buried treasures, the outposts of the universe have beckoned to the adventurous from immemorial times."[1] In

1. The Possibilities of Large Telescopes, *Harper's Magazine* (April 1928).

addition to his remarkable ability to raise funds for large telescopes, Hale was a great scientist who applied physics and ingenuity to the study of the sun and stars. As a 22-year-old undergraduate at the Massachusetts Institute of Technology, he constructed the first spectroheliograph. By allowing the sun's disc to drift across the entrance aperture of a spectrograph, this device can create an image of the sun at just one wavelength. The invention enabled Hale to undertake pioneering studies of solar activity using a telescope at his home in Chicago funded by his father (whose fortune was made in providing elevators for skyscrapers). It brought him immediate international fame. Within 2 years he had secured a position at the newly founded University of Chicago and procured funding for a 40-inch refractor with an objective lens weighing 225 kilograms—the largest in the world. The observatory was named after the donor, Charles Yerkes, a streetcar millionaire who, incidentally, played a significant role in establishing the London Underground. Thus began a century of astronomical discovery at the University of Chicago.

Hale was an intensely energetic and ambitious man. His motto "Make No Small Plans" led to frequent bouts of exhaustion, nervous breakdowns, and essential recuperative vacations away from the stresses of his tempestuous life. He referred to his medical condition as "Americanitis." Not content with building the largest refracting telescope at Yerkes Observatory, he set upon a lifetime's path to raise funds for, to construct, and to oversee the operation of a succession of giant reflecting telescopes, each one the world's largest at the time of its completion. Moving to Pasadena in 1904 as founding director of the Mount Wilson Observatories, he raised funds for a 60-inch reflector from the recently formed Carnegie Institution of Washington and, even before its completion, secured a donation for the more powerful 100-inch telescope from John D. Hooker, a hardware magnate, which was completed in 1917. For his newly established observatory, Hale hired an army of talented observers, including Harlow Shapley (later head of Harvard College Observatory) and Edwin Hubble, making Pasadena the centre of world astronomy. So began the exploration of the distant universe.

At a time when many observational astronomers were preoccupied studying individual stars, Hale was amongst the first to realise that large telescopes had the power to explore the vast reaches of space, and that, with the finite velocity of light, they could look back in time. When Shapley used the 60-inch to estimate that a globular star cluster was 36,000 light years away, Hale immediately recognised the implications of looking back into cosmic history. Shapley further conjectured such star clusters belonged to a disc-like Milky Way that could be 300,000 light years across. Our knowledge of the size of the visible universe increased further when, in 1923, Hubble used the 100-inch Hooker Telescope to locate Cepheid variable stars (chapter 2) in the Andromeda spiral, and established that it was 850,000 light years away. This meant it lay outside the confines of the Milky Way altogether.[2] The Milky Way, previously considered to be the entire universe itself, was just one of many billions of galaxies populating an unforeseen immensity of space. This discovery was simply fantastic and opened the door to unimagined discoveries.

Contemplating his next step, Hale claimed in the magazine *Popular Astronomy* that the scientific achievements of his 60-inch and 100-inch reflectors "promise well for larger apertures."[3] Five years later, in *Harper's Magazine*, he wrote that "Starlight is falling on every square mile of the Earth's surface and the best we can do is to gather up and concentrate the rays that strike an area 100 inches in diameter."[4] He was making the case for a yet larger reflector. Securing $6 million from the Rockefeller Foundation (equivalent to $100 million today), the largest donation ever at that time for a single scientific endeavour, the story of the construction of the 200-inch (5-metre) mirror and telescope for Mount Palomar is a classic adventure story with numerous seemingly unsurmountable hurdles, each one eventually overcome (see plate 11).[5]

2. Modern distance estimates place the Andromeda spiral even further away, at 2.5 million light years.

3. The Possibilities of Instrumental Development, G. E. Hale, *Popular Astronomy*, vol 31 (1923), p573.

4. The Possibilities of Large Telescopes, *Harper's Magazine* (April 1928).

5. I have named this chapter after the splendid story of the Hale 200-inch telescope told authoritatively by Ronald Florence in *The Perfect Machine* (Harper & Collins 1994). Also worth watching is the PBS documentary *The Journey to Palomar* produced by Todd and Robin Mason.

The Palomar telescope was later named after George Ellery Hale, even though he never lived to see its completion, having died at age 69 in 1938. Quite apart from being the most successful optical telescope for at least the next 40 years, it championed many technical innovations. The primary mirror is made of Pyrex, a low-expansion material introduced by Corning, a glassware company in New York state. The telescope structure has a horseshoe-shaped mount and sits on oil-pressure bearings. It incorporates a differential flexure system for maintaining precise optical alignment as the telescope traverses the sky. Compared with the 60- and 100-inch telescopes, it has a faster (shorter focal length) f/3.3 primary mirror and rapidly interchangeable focal stations. Twenty years in the making, through the Great Depression and Second World War, it represented the most famous scientific instrument of its time. Alongside air travel, radio and television broadcasts, and the movie industry, it encapsulated the American technical revolution of the twentieth century.

Exploring the distant universe did not stop while Pasadena astronomers awaited the completion of the 200-inch at Mount Palomar. Edwin Hubble continued to push the frontiers using both the 60- and 100-inch telescopes. Hired in 1919 after reaching the rank of major in the First World War, Hubble retained an affected English accent from his pre-war time as a Rhodes Scholar studying law at Oxford university. His work in the early 1920s on the morphological classification and three-dimensional forms of external galaxies and on the nature of gaseous nebulae within the Milky Way, and his demonstration of the distances to nearby spirals using variable stars, set the stage for an astonishing decade of discovery and the foundation of modern observational cosmology.

Hubble's early work that led to the idea of an expanding universe using Vesto Slipher's velocities and his own distances from variable stars is discussed in chapter 2. Although often credited as discoverer of the "expanding universe," he was undecided whether the recessional velocities were real or due to some unknown law of nature in a static universe. Unusually for Hubble, who was very much a loner, he teamed up with Milton Humason (1891–1972), a high-school dropout, formerly a mule

driver and janitor on Mount Wilson, to measure galaxy velocities at
greater distances using the 100-inch telescope.[6] The goal was to extend
the velocity-distance relation to greater depths, where any deviations
from the local linear (i.e., straight-line) relation might test the physical
origin of these enormous velocities. Whereas by 1925 Slipher had used his
spectrograph at Lowell Observatory in Flagstaff to secure 42 velocities,
with the largest at 1800 km/s, by 1936 Humason had used the 100-inch
to measure velocities for a further 150 galaxies, with the largest at
42,000 km/s, corresponding to a redshift of $z = 0.14$.

At such vast distances (around 2 billion light years, using present
cosmological models), individual variable stars could not be discerned,
so Hubble and Humason had to rely on the total brightnesses of their
galaxies as a distance indicator, assuming they shared similar intrinsic
luminosities. Their velocity-distance relation was, in actuality, simply
a velocity-brightness diagram, since if all galaxies were equally luminous their distances would be directly related to how faint they were.
Although this assumption was incorrect, because galaxies span a wide
range of luminosities, the meticulous details in Hubble and Humason's
1931–1936 articles laid the framework for later cosmological studies. To
test whether the velocities were due to the expansion of the universe,
they corrected the brightness of each galaxy for two effects: an "energy
effect" (which accounts for the redshift of light rays) and a "number
effect" (which accounts for the cosmic expansion of the path length
that reduces the arrival rate of light photons). By 1936, both observers
had exhausted the capabilities of the 100-inch and were awaiting completion of the 200-inch. Hubble estimated that extending the survey to
galaxies with velocities of over 60,000 km/s would determine whether
these two effects could be confirmed or rejected, and hence finally determine whether the cosmic expansion was real. As late as 1953, just

6. Humason is referred to as an "affable, imprecationist, rake and rogue, gentleman and
friend" by Allan Sandage in his *Centennial History of the Carnegie Institute of Washington*, vol 1,
The Mount Wilson Observatory (Cambridge 2004), p496. A rabid Republican in charge of telescope scheduling, Humason ensured Democrats on the observatory's staff were allocated observing time in early November and thereby unable to vote in presidential elections.

before his fatal heart attack, Hubble remained unconvinced that the universe was expanding.[7]

Working with Caltech's theoretical cosmologist Richard Tolman, Hubble also pioneered the technique and interpretation of counts of faint galaxies. Visually inspecting an archive of over 1200 photographic plates taken with the 60- and 100-inch telescopes, Hubble sorted over 44,000 galaxies as a function of their apparent brightness. If the faintest galaxies were more distant, their numbers should increase proportionally with the larger volume of space being explored. In the familiar Euclidean space of everyday life, volume increases as the cube of the radial distance. However, Einstein's theory of relativity permits that three-dimensional space can be positively or negatively curved. Just as the large distances between points on the Earth's curved surface, and the angles these distances subtend when joined to form triangles, do not obey the laws of geometry on a flat piece of paper, so a positively curved universe will reveal a smaller volume for a given radial distance, and a negatively curved universe will have a larger volume and show more galaxies. According to the general theory of relativity, the curvature of space is governed by the amount of cosmic matter, which, in turn, also controls the fate of the expansion. For example, in a dense universe, space is positively curved and the expansion is slowed down by the gravitational attraction of matter, in much the same way as the motion of a ball launched upwards is retarded by the Earth's gravity. In 1936 Hubble demonstrated that the number of faint galaxies was smaller than expected for a static Euclidean universe. However, making the corrections for the effects expected in an expanding universe discussed earlier, his galaxy counts yielded such a positively curved universe that its age and size was improbably small. This forced him again to doubt that the velocities he and Humason had painstakingly measured were due to a cosmic expansion. Although we now understand the inaccuracies in Hubble's corrections

7. On May 8, 1953, Hubble delivered the Darwin Lecture of the Royal Astronomical Society, entitled "The Law of Red-Shifts," which summarised his momentous observational achievements and personal conclusions. He died on September 28 that year, and the lecture was published posthumously without correction in the *Monthly Notices of the Royal Astronomical Society*, vol 113, p658.

and the fallacy of this analysis, his 1936 paper represented the final word on this important topic of faint-galaxy counts until the 1970s.

The Palomar 5-metre (200-inch) telescope was finally dedicated and named after Hale at a gala event on June 3, 1948, with his widow, Evelina, on the dais. A condition of the Rockefeller grant was that the observatory would belong to the California Institute of Technology (Caltech), a private university in Pasadena, but operations would be shared with the Mount Wilson Observatories. This condition surprised both the director of the Mount Wilson Observatories, Walter Adams (who took over from Hale in 1923 as a result of the latter's medical difficulties), and John Merriam, a palaeontologist who was president of the Carnegie Institution of Washington (responsible for funding and oversight of the Mount Wilson Observatories). Although Caltech engineers certainly played the leading role in designing and assembling the Palomar telescope, and polishing its primary mirror to perfection during the 1930s and early 1940s, the university did not have a well-established astronomy department. The Rockefeller Foundation's condition led to a forced marriage and a continued rivalry between these two Pasadena institutions that lasted well into the next century, as I witnessed after I moved to Caltech in 1999. Recognising the urgent need to properly exploit its massive new telescope, in 1948 Caltech hired Jesse Greenstein (1909–2002) from Chicago to set up a new astronomy programme. One of its inaugural graduate students was Allan Sandage (1926–2010), who later picked up the baton from an ailing Hubble and continued the great man's work, now using the 200-inch telescope.

I first met Sandage at a conference in Hawaii in 1986, and in 1992 I invited him to give a prestigious lecture at Durham University, where we got on pretty well (plate 12). He had also decided on an astronomical career at a very early age and built his own telescope. Aged 8, he decided that he *had* to become an astronomer because he was "Compelled. Out of a sense, not so much of duty, but there was just nothing else that seemed as worthwhile."[8]

---

8. Interview with Alan Lightman and Roberta Brawer in their *Lives and Worlds of Modern Cosmologists* (Harvard 1990), p69.

Hubble's final years must have been ones of great personal disappointment. After waiting for the arrival of the 200-inch for nearly two decades, he was offended not to be offered the directorship of the newly enlarged Mount Wilson and Palomar Observatories when Walter Adams retired in 1946. The decision most likely reflected both Hubble's lack of administrative experience and an aloofness towards his fellow astronomers. Indeed, he could be surprisingly arrogant with colleagues and competitors. Instead, the directorship passed to Ira Bowen, a laboratory spectroscopist and very tactful leader who made important technical contributions to improve the performance of the new telescope. Then, with bad timing, Hubble suffered his first heart attack in 1949 and was forbidden by his doctor to observe for a year. Hubble and Humason therefore gave Sandage important responsibilities within their programme, initially at Mount Wilson and later at Palomar. They soon realised Sandage's qualities as a meticulous observer and committed researcher. As a result, Bowen hired the young Caltech graduate as a staff member in 1952.

At this point Sandage knew he had a profound duty to continue Hubble's research on the cosmic distance scale. In a statistic that will astonish today's astronomers, only 12 staff members were permitted to use the newly completed telescope, the largest in the world, and none of them was experienced in studies of faint galaxies. Moreover, no other telescope was up to the task Hubble had defined. Humason expressed his eagerness to work with Sandage and urged him to be leader of the continuing programme. In an interview, Sandage confessed that the quest "was all laid out and I was the only one left after Hubble died."[9]

Although Sandage agreed to continue Hubble's programme, he was aware of two issues. The first was that Hubble regarded his observational programme as an empirical adventure without much connection to cosmological theory. Hubble foresaw extending the recessional velocity–apparent brightness diagram to larger distances largely as a means of detecting or rejecting some deviation from a linear relation, perhaps to refute the cosmological expansion. Sandage realised the

9. Quote from *Lonely Hearts of the Cosmos* by Denis Overbye (Harper Collins 1991), p29.

need for a more scientific approach based upon testing specific cosmological models. This meant he had to master the relevant mathematics. The second issue related to a complaint he frequently heard from a senior colleague, Walter Baade, whom he greatly respected. Baade (1893–1960), a stellar spectroscopist, had escaped Nazi Germany and arrived at the Mount Wilson Observatories in the 1930s. His work led to the definition of two distinct types of stellar populations in galaxies (termed Populations I and II), reflecting their different ages and chemical compositions. The distinction affected the luminosities of some of the variable stars Hubble had used to estimate his cosmological distances. In 1952, just as Sandage was completing his PhD thesis with Baade, who had taken over as his formal supervisor owing to Hubble's convalescence, Baade estimated the size of the universe to be fully twice that determined by Hubble. In a prescient remark that reshaped the field 20 years later, Baade emphasised to Sandage that "You must understand the physics of the galaxies before you can use them to mark the geometry of space."[10]

In 1961, Sandage published his observational strategy for the 200-inch telescope in an article entitled "The Ability of the 200-inch Telescope to Discriminate between Selected World Models," which remains one of his most cited papers. Of the four tests discussed, he considered the most promising to be the "magnitude-redshift (m-z) relation," where the magnitude m represents the apparent brightness of a galaxy. If it were possible to reach a redshift $z \sim 0.5$ with the 200-inch, he claimed, it would be possible to discriminate between a "closed" universe that would stop expanding and an "open" one that would expand forever. He claimed that merely counting galaxies as a function of increasing faintness, another test that Hubble had pioneered, would be insufficiently sensitive for the task. For the m-z test, Sandage proposed using the most luminous galaxies in clusters. Not only would these be easier to detect at high redshift, but they seemed more likely to be of fixed luminosity. However, he warned the path would be arduous with many possible

---

10. Allan Sandage, *Centennial History of the Carnegie Institute of Washington*, vol I, *The Mount Wilson Observatory* (Cambridge 2004), p508.

pitfalls in interpretation. His 1961 article is a masterpiece, detailing as it does the careful plans of a great scientist.[11]

Sandage considered the nature and fate of the expanding universe to be governed by two numbers—the Hubble constant H_0 and a *deceleration parameter* q_0. The subscripts "0" here refer to the present-day value of these two quantities. H_0 measures the current rate of the cosmic expansion; in other words, how rapidly velocities increase at greater distances. q_0 determines by how much the rate of expansion is *slowing down* as time progresses; for example, owing to gravitational attraction because of the presence of matter. In terms of the *m-z* relation, q_0 would measure how the relation departs from a constant trend, a straight-line or linear relation, as one pushes to higher *z*. Whereas Hubble and Humason had been empirically motivated to find such a departure from a straight line, Sandage now fully grasped the significance in terms of detailed cosmological models. A clear measurement of q_0 would also determine the fate of the universe. If $q_0 \geq 0.5$, it would signify a closed universe that would eventually stop expanding owing to the presence of a sufficient amount of matter; otherwise we would live in an open universe that would expand forever.

After Hubble's death, Sandage worked, initially with Humason until he retired in 1957, to secure the redshifts and brightnesses of galaxies in clusters out to redshifts of $z \sim 0.5$. However, improved detector technology made a major contribution in later years. James "Jim" Westphal, a Caltech professor who was later to become director of Palomar, worked with Jerome Kristian at the Mount Wilson and Palomar Observatories in the development and exploitation of a silicon-intensified target camera detector that was 20 times more sensitive than the photographic plate. Sandage worked with this Caltech-Carnegie[12] duo to extend his galaxy studies to a redshift $z \sim 0.75$. The scope of the overall Palomar campaign was unprecedented. Sandage clocked up several hundred

11. A. Sandage, *Astrophysical Journal*, vol 133 (1961), pp355–92.

12. The Mount Wilson Observatories changed to become Mount Wilson and Palomar Observatories when the 200-inch was completed in 1948. In 1980 it became the Carnegie Observatories, reflecting its funding from the Carnegie Institute of Washington.

nights of observing, which led to two series of monumental scientific articles: eight papers as sole author between 1972 and 1975 and two papers between 1976 and 1978 by Kristian, Sandage, and Westphal.

Before discussing the results of Sandage's heroic observations, I will digress and describe an astonishing discovery that overshadowed the above programme and, at one point, so upset Sandage that he offered to resign from Mount Wilson and Palomar Observatories. By 1960, the most distant known object confirmed at Palomar was the radio galaxy 3C295 at a redshift $z = 0.46$. Although the spectrum was secured by Rudolph Minkowski, a Mount Wilson and Palomar astronomer who specialised in studies of radio sources, Sandage was instrumental in this success by taking the relevant photographs beforehand that provided an accurate position for the faint galaxy and enabled Minkowski to acquire its spectrum. Not only was 3C295 the most distant known object but it was remarkably compact and stellar in appearance. With another Caltech radio astronomer, Tom Matthews, Sandage began using the 200-inch to follow up other compact radio sources from the third Cambridge radio catalogue (hence the designation "3C"). The visual appearance of these compact radio sources led to the designation "quasi-stellar radio source," which was soon shortened to "quasar." A second source, 3C48, was remarkably blue, varied in its brightness, and had a spectrum unlike any normal star. Sandage showed the curious spectrum to both Ira Bowen and Jesse Greenstein, but neither was able to decipher it. Variability was key to limiting the size of this mysterious source. It would surely require some supernatural coordination for an entire galaxy to vary its brightness on human timescales, so Sandage imagined 3C48 to be a very unusual "radio star," distinct from 3C295; he delayed publication to get more data, hoping to better understand its physical nature. While more quasars were found in the 3C radio list, nobody knew what they were.

In the meantime, however, a Dutch radio astronomer, Maarten Schmidt, arrived at Caltech and was given an accurate position for a much brighter Cambridge radio source, 3C273. Although its spectrum was also puzzling, with Jesse Greenstein's help it was eventually determined to be at a redshift $z = 0.16$. The inferred luminosity was astonishing: for such a bright object to be so distant, it would have to be 100 times brighter than even the most luminous galaxy. With this insight, Greenstein then realised that Sandage's spectrum of 3C48 indicated a redshift $z = 0.37$. In 1963 Schmidt and Greenstein each rushed an article into the journal *Nature*, announcing quasars were a new, extraordinarily luminous constituent of the universe. Even though these quasars were amongst the most distant sources known, to be consistent with their variability their intense luminosity had to originate in a volume the size of the solar system. Their existence seemed to defy the laws of physics.

We now know quasars represent intensively active regions in the compact nuclei of certain galaxies. Their astonishing luminosity does not originate in starlight but in non-thermal radiation that arises when electrically charged gaseous material spirals into a super-massive black hole. The black hole in 3C273, now known to be the brightest quasar in the sky, has a mass 900 million times that of the sun. If 3C273 were placed alongside our nearest stars, it would outshine the sun.

For Sandage, the Caltech discovery was a bitter blow; he could rightly claim to be the discoverer of quasars and their variability but was omitted from two well-publicised articles that announced their extragalactic nature and spectacular luminosities. Although frustrated by losing the glory, Sandage was keen to move forward by exploiting quasars for cosmology. Such luminous objects could be seen to enormous cosmic distances and hence perhaps be used to determine q_0, the measure of how fast the cosmic expansion is slowing down. Pretty soon, quasars were being detected at redshifts as high as $z = 2$, well beyond any galaxies at the time. Although the number of high-redshift quasars was still modest, Sandage was already examining their magnitude-redshift (m-z) relation. Moreover, noting their blue colours, Sandage finally stumbled by chance on a second discovery. Using wide-field photographic plates of his 3C radio sources, he noticed many other starlike sources of similar colours in the same field of view. He wondered whether for every radio-loud quasar, such as those first located using the 3C catalogue, there was a yet larger population of *radio-quiet* quasars with similar blue colours. After demonstrating his idea by confirming spectroscopically that a few radio-quiet sources were also at great distances, he was able to respond to Schmidt and Greenstein's discovery with his own. His 1965 paper was aimed to match theirs in significance and was entitled "The Existence of a Major New Constituent of the Universe: The Quasi-stellar Galaxies."[13]

Unfortunately, Sandage was so keen to exploit this sample of radio-quiet quasars before anyone else, that he could not resist analysing their numbers to determine the value of q_0. Since he hadn't had time to study all of these new-found radio-quiet blue sources spectroscopically, he made the rash assumption that all were distant quasars. This turned out not to be the case: some were stars in the Milky Way. Although Sandage had indeed located a new population of radio-quiet quasars, he had significantly overestimated their numbers. Moreover, he had to infer distances (and hence redshifts and associated volumes) simply from their apparent brightness, assuming quasars were "standard candles"—that is, they shared identical luminosities. For this assumption, there was no convincing evidence. By Sandage's standards, the article was hurried and many experts dismissed its conclusions.[14] Sandage himself was embarrassed and offered his resignation to the Mount Wilson and Palomar Observatories' director, Ira Bowen (who declined to accept it).

Returning now to his study of brightest cluster galaxies, in 1972 Sandage present an *m-z* relation based on 84 clusters. The tightness of this relation convinced him, at the time, that the brightest cluster galaxy was an excellent "standard candle" for estimating distances on cosmic scales (figure 3.1, *left*). He derived a value for the deceleration parameter $q_0 \approx 1$ that, while uncertain, was consistent with a closed universe and the ages of the oldest star clusters. While discussing a growing concern (which dated back to Baade's premonition in 1952) that galaxies may have been more luminous in the past owing to the effects of stellar evolution, he considered such an "evolutionary correction" to be too uncertain to include. For his results to be consistent with an open universe,

---

13. A. Sandage, *Astrophysical Journal*, vol 141 (1965), pp1560–78.

14. A further feature of the article that annoyed some astronomers was that it was accepted for publication on the date of submission, implying the journal editor did not think such a major article from Sandage needed any external refereeing.

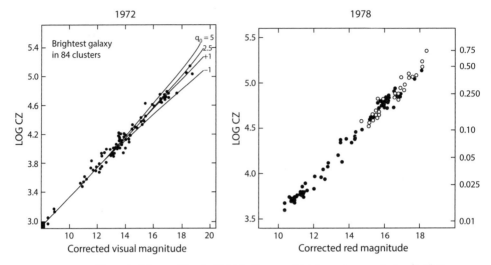

FIGURE 3.1. The end of the road for the Hubble diagram of brightest cluster galaxies. (*Left*) By 1972, Sandage had extended the linear diagram (here shown as redshift times the speed of light, CZ, versus a corrected apparent magnitude, a logarithmic measure of brightness with larger numbers for fainter sources) to a redshift $z \sim 0.5$, claiming a deceleration parameter $q_0 \sim 1$. (*Right*) By 1978, with Kristian and Westphal, the diagram was extended to $z = 0.75$, where a first hint of an upturn in the relation at high redshift was claimed with a formal value of $q_0 \sim 1.5$. Such a universe would be closed and stop expanding in the future, but its present age would be less than those of the oldest stars. The conflict pointed to a possible fading of galaxy luminosities over the past several billion years and, sadly, the fundamental limit of the method as a way of determining the world model.

he believed it was "unlikely that [galaxies] fade by [the required] amount"[15] and questioned even if the sign of the (evolutionary correction) was known. Although many believed the collective luminosity of stars in a galaxy should fade with time owing to their nuclear fuel running out, others argued the brightest galaxies in clusters might merge with their neighbours, thereby growing more massive and luminous with time. The two effects could easily cancel each other out, in which case there would be no evolutionary correction.

Nonetheless, the implications for the nature and fate of the universe remained highly uncertain. After this enormous observational effort

15. The Redshift-Distance Relation. II. The Hubble Diagram and Its Scatter for First-Ranked Cluster Galaxies: A Formal Value for q_0, A. Sandage, *Astrophysical Journal*, vol 178 (1972), p22.

over nearly 20 years, most astronomers today would wonder whether continuing the campaign was worth it. Whether because of his continued devotion to Hubble's original plans or anticipation of the promise of new detectors, Sandage decided to soldier on. From 1975 to 1977, with Kristian and Westphal, redshifts and photometry were obtained for 50 more clusters, using both the spectrograph with the silicon-intensified target camera and traditional photomultipliers. This progress included the heroic measurement of a redshift $z = 0.75$ for the radio galaxy 3C343.1. In their 1978 paper, the authors claimed for the first time that they had detected curvature in the m-z relation, with a value of the deceleration parameter even larger than in 1972 (figure 3.1, *right*). This implied, at face value, a universe that was definitely closed but somewhat younger than the age of the oldest star clusters. A possible reconciliation of the results with independent measurement of the low mass density of the universe (implying an open model) is briefly discussed, but the paper ends limply without a clear conclusion.[16]

By the early 1970s, Jesse Greenstein had built a world-class astronomy department at Caltech. Recognising the unique potential of the 200-inch, there were newly hired observers, theorists as well as innovative instrumentalists. One rare individual, James "Jim" Gunn, is skilled in all three aspects of astronomical research. A Caltech graduate whose PhD was in theoretical cosmology, he returned from Princeton to join the faculty from 1970 to 1980 and built several innovative instruments to make pioneering observations at Palomar. In the early 1970s he teamed up with John Beverly "Bev" Oke (1928–2004), a modest Canadian at Caltech, who had developed a multichannel photometer that could scan the energy spectrum of a galaxy extremely efficiently. The pair did the unthinkable and set out to challenge Sandage's historic right to measure the cosmic distance scale (see plate 13).

Sandage never revealed to me personally how he viewed this competition from the Caltech rivals, but it is claimed he tried to prevent Gunn

---

16. The Extension of the Hubble Diagram. II. New Redshifts and Photometry of Very Distant Galaxy Clusters: First Indication of a Deviation of the Hubble Diagram from a Straight Line, J. Kristian, A. Sandage, and J. A. Westphal, *Astrophysical Journal*, vol 221 (1978), pp383–94.

and Oke getting observing time and treading on his turf. Apparently when asked who was the best astronomer, Sandage replied, "Well, young Jim Gunn is doing pretty well. If he keeps it up he could be number two."[17] While Gunn and Oke generously acknowledged Sandage's wisdom and input in a 1975 article, their observational strategy towards estimating the deceleration parameter differed significantly from Sandage's. Oke's multichannel photometer was capable of frequently switching from a faint galaxy to the adjacent night sky and, for 28 clusters beyond a redshift $z \sim 0.1$, it was used efficiently to measure the energy distribution of each of the brightest cluster galaxies. This approach bypassed the "energy" correction (sometimes called the K-term) that plagued earlier work since the brightness of each galaxy could now be calculated at the same "rest-frame" frequency—that is, as if the galaxy had no recessional velocity. The downside of this laborious observational campaign was a smaller sample that did not reach beyond a redshift of $z = 0.46$. Their value for the deceleration parameter $q_0 = 0.3$ was much less than Sandage's 1972 estimate of $q_0 \approx 1$, perhaps suggesting an open universe that expands forever.

However, the most significant development arose from Gunn's interaction with a theorist, Beatrice Tinsley, who was modelling the evolution of galaxies on a computer. Visiting Caltech in 1972, Tinsley convinced Gunn that beyond redshift $z \sim 0.5$, the types of galaxies found in rich clusters would have been much more luminous owing to the effects of stellar evolution. Although Gunn and Oke declared in their 1975 paper that their work was continuing with more distant clusters, on the basis of Tinsley's calculations the article included the prophetic admission "we find that uncertainties in the evolution are comparable to statistical uncertainties at present, but with very much fainter samples they will completely dominate."[18] After 1975, Gunn abandoned the quest for the deceleration parameter and used his Palomar time to focus on the evolution of galaxies.

17. Quote from *Lonely Hearts of the Cosmos* by Denis Overbye (Harper Collins 1991), p178.

18. Spectrophotometry of Faint Cluster Galaxies and the Hubble Diagram: An Approach to Cosmology, J. E. Gunn and J. B. Oke, *Astrophysical Journal*, vol 195 (1975), p255.

Beatrice Tinsley (1941–1981), a forthright young New Zealander who completed her PhD thesis at the University of Texas in Austin in 1966, argued that most of the stars in the cluster galaxies being targeted by Sandage and others formed a long time ago in a single burst of star formation. Accordingly, as they aged, she argued, the more massive and luminous stars would run out of their nuclear fuel first and eventually fade away. Such galaxies would therefore dim monotonically as the universe grew older. Running time backwards, the most distant galaxies would be more luminous than those today, causing the upturn in the *m-z* relation that Sandage incorrectly claimed was due to a closed universe. Tinsley showed the effect was large enough to change the conclusion of Sandage's beloved *m-z* relation; the universe must be open and would expand forever. Such a result was also consistent with a growing number of estimates of the low density of gravitating material in the universe. Following a paper led by Tinsley that amassed the emerging evidence for an open universe in 1974, apparently one media report claimed "Caltech astronomers declare the universe is now open!"

Sometimes I wish I had been born a decade earlier so that I could have participated in the fundamental debate, championed by Tinsley, that heralded a major change in direction in the studies of distant galaxies—from determining the nature of the universe in which we live to tracing the evolution of stars in galaxies back to their birth, the theme of this book. It seems odd that Sandage, who in the 1960s pioneered the technique of age-dating star clusters based on a physical understanding of how the more massive stars die first, would hold out against growing evidence for the dominant effect of galaxy evolution. Hubble's vision of using distant galaxies to determine the fate of the universe, a campaign pursued over half a century with the 100- and 200-inch telescopes, would lead to a conclusion identical to Baade's warning to Sandage in 1952—namely, that one has to understand galaxies before using them for cosmology.

I didn't meet Beatrice Tinsley until my first visit to the United States in 1977 as a young postdoctoral researcher. She gave me her evolutionary models for interpreting my number counts of faint galaxies. I first met Jim Gunn in London in 1979 at a meeting discussing the relatively new field of galaxy evolution. His international reputation was already formidable but I found him very approachable. When I first met Allan Sandage in 1986 he was welcoming and friendly, and apparently interested in my work—although at the time I was not quite sure if I was just being humoured. By then, however, the arguments were largely over. Even Sandage agreed we lived in a universe that will expand forever. The future looked dismal and lonely. Galaxies would get further apart and continue to dim as more and more of their stars died. The role of our large telescopes was to look back to trace the glorious past of the population of galaxies, if possible back to the spectacular birth of galaxies from the primordial darkness—cosmic dawn![19]

A decade later, two teams of astronomers resurrected Sandage's m-z relation as a cosmological test, this time using a certain class of supernovae. Such exploding stars can, briefly for a few days, outshine an entire galaxy, and thus can be seen to high redshifts. However, unlike brightest cluster galaxies, studies of nearby supernovae of this type indicate that they reach a near-constant maximum luminosity and so can serve as reliable "standard candles" for estimating cosmological distances. The two teams, one led by Saul Perlmutter at Berkeley (of which I was a founding member) and the other led by Brian Schmidt, then at Harvard, set out to measure the parameter q_0, which, recall, indicates the rate at which the expansion of the universe is slowed down by the gravitational attraction arising from its matter content. To both teams' astonishment, their results, published in 1999, indicated that the cosmic expansion is not slowing down but speeding up; we live in an accelerating universe! Although there were earlier suggestions of a missing cosmic energy source from the large-scale clustering of galaxies and the assumption that three-dimensional space is flat (or Euclidean), the result was still a big surprise as there is no obvious physical explanation for this acceleration. It is as if, in addition to the attractive force of gravity, there is a repulsive aspect of empty space, a feature that has been given the moniker "dark energy." Understanding the nature of this

---

19. A sympathetic, but perhaps somewhat unbalanced, biography of Beatrice Tinsley is *Bright Star: Beatrice Hill Tinsley, Astronomer* by C. C. Catley (Cape Catley 2006).

mysterious dark energy represents one of the most outstanding questions in current cosmology. Perlmutter, Schmidt, and Adam Riess, a prominent member of Schmidt's team, shared the 2011 Nobel Prize in Physics for this unexpected discovery.[20]

To return to the previous story of Californian "cosmological theatre" (in which of course I played no part), I now fast-forward to when I arrived at Caltech in 1999 (chapter 7). Within 6 months of my arrival, I was appointed director of Palomar Observatory. This was a curious situation for me personally. As I recount in the next chapter, I had spent much of my career in studies of galaxy evolution competing with the Californians, who had privileged access to the Hale 200-inch, yet now I was responsible for managing this venerable facility and overseeing its scientific output. I took over the mantle from Wallace "Wal" Sargent—a fellow Brit originally from Lincolnshire—who pioneered the study of the intergalactic medium via quasar absorption lines, the very topic of my undergraduate research project in 1968. It seemed such a coincidence. Wal once said that when he first arrived at Caltech and began observing at Palomar, he was initially scared when he encountered the massive 200-inch. He would feel a "slight pit in his stomach" when entering the observing floor, as if the science he was doing "wouldn't be good enough for such a grand machine."[21] Despite the development of larger facilities—most notably the twin 10-metre Keck telescopes, each of whose light-gathering power is four times that of the Hale Telescope— the "Perfect Machine" continues to do groundbreaking science more than 70 years after its completion in 1948 (plate 14).

20. A vivid "insider" account of the discovery of the accelerating universe is *The Extravagant Universe: Exploding Stars, Dark Energy, and the Accelerating Cosmos* by R. P. Kirshner (Princeton University Press 2016).

21. Quote from interview in *Star Men* (Canada Media Fund 2015; © Inigo Athenaeum Enterprise Inc.), a documentary film directed by Alison Rose.

4

The Anglo-Australian Revolution

The first Australian astronomer I met was the Perren Professor of Astronomy at University College London, C. W. Allen. In early 1968, prior to my admission to UCL, he interviewed me in his small office at the Mill Hill Observatory and, apart from asking me to solve some mathematical equations, the occasion was remarkably relaxed. As he taught only students in the final year of the undergraduate programme, "the professor" was rarely encountered and was regarded with considerable respect. To us young students, he epitomised the caricature of an eccentric academic. One day he marched into the lecture room at Mill Hill, quickly filled the blackboard with equations and then turned to face us. He paused and, realising he was about to give a lecture to the wrong class, promptly marched out without saying a word. We witnessed many other amusing incidents. Once we saw him remove his overcoat on leaving the building and then, on re-entry, put it back on. As a mischievous upstart, I frequently imitated his behaviour to the amusement of my fellow students.

Clabon Walter Allen (known to his colleagues as "Clay") was famous for authoring a compendium of astronomical data called *Astrophysical Quantities*, a reference book that, until the internet age, could be found on the desk of every self-respecting astronomer. He was a distinguished solar astronomer, which, he once commented, allowed him to be a "daybird." When he taught us solar physics in the final year, he also reminisced about his wartime duties for the Australian government. Although most people would regard the sun as an unchanging and thankfully reliable source of energy, it does vary in its output, with occasional high-energy flares that cause havoc with long-distance radio transmission. During the war, Allen's principal task was to monitor sunspots on the sun's surface in order to predict when a solar flare might disrupt the military's long-distance radio transmissions. With a twinkle in his eye, he informed us, "I didn't get a single one right!"

In my final undergraduate year, I was revising by looking over previous examination papers. One question, set by Professor Allen, mentioned the "3.9 metre Anglo-Australian Telescope (AAT) now under construction." This greatly interested me. Despite attending regular meetings of the Royal Astronomical Society, I hadn't heard of this project. Although it occurred to me that a 3.9-metre (150-inch) telescope would mean a huge increase in observing capability for the UK community, I never imagined that it would totally transform British observational astronomy and, in turn, establish my own career.

This partnership in optical astronomy between the United Kingdom and Australia originated in a proposal by Richard van der Riet Woolley, the director of the Mount Stromlo Observatory in Canberra (he later became director of the Royal Greenwich Observatory, the UK Astronomer Royal, and Sir Richard Woolley; see chapter 1). As early as 1953, Woolley proposed a 200-inch Commonwealth Southern Observatory as a southern-hemisphere counterpart to Caltech's recently completed Palomar Observatory in southern California. His successor at Mount Stromlo, the Dutch American astronomer Bart Bok, strongly supported the idea and, in 1957, commissioned a survey for an appropriate site. Siding Spring, a mountain in the volcanic Warrumbungle range near the small town of Coonabarabran, New South Wales, was selected in 1962.

Astronomers consider the southern celestial sky to be more glorious than its northern counterpart because it hosts the central regions of our Milky Way galaxy and offers unique access to the twin Magellanic Clouds, satellite galaxies that are 10 times closer than the Andromeda spiral. By 1953, both the United Kingdom and Australia had already established some government-run southern-hemisphere facilities. The

Radcliffe 74-inch reflector in South Africa was completed in 1948, following the decision to transfer an observatory originally funded at Oxford by the Radcliffe Trustees, who were charged with managing the estate of John Radcliffe, MD (1650–1714), the most notable physician of his day. The telescope in Oxford prior to this decision was none other than the one I used as an undergraduate at UCL's Mill Hill Observatory (chapter 1); it was relocated in 1938. In 1948, the new Radcliffe 74-inch reflector became the largest optical telescope in the southern hemisphere and served both British and South African astronomers. The Australians had established the Commonwealth Solar Observatory on Mount Stromlo in 1924, and later installed a similar 74-inch reflector in 1955.[1] Both countries also pioneered the new field of radio astronomy after the Second World War, with the 250-foot Jodrell Bank and 210-foot Parkes telescopes being particularly influential in their discoveries.

In 1965, 3 years after selecting the Siding Spring site, Woolley and Bok, in their respective leadership positions, proposed an Anglo-Australian 150-inch telescope to the UK and Australian governments at a cost (by 1967) of 11 million Australian dollars and an annual operating budget of A$450,000. The proposal was accompanied by letters of endorsement from the Royal Society of London and the Australian Academy of Science. The partnership was not an obvious one given the geographical distance between the two astronomical communities, and, indeed, the route to completion would be fraught with misunderstandings and disagreements.[2] Some UK astronomers proposed expanding the facilities in South Africa, while the British government would have preferred their astronomers to join a fledgling European effort that became the European Southern Observatory (ESO; chapter 9). Many Australian astronomers wished to partner with the more experienced Americans, but by the mid-1960s the United States had already developed its own plans in Chile.

1. *Mt Stromlo Observatory: From Bush Observatory to the Nobel Prize*, R. Bhathal, R. Sutherland, and H. Butcher (CSIRO Publishing 2013).

2. A faithful account of the difficulties is given in *The Creation of the Anglo-Australian Observatory* by S.C.B. Gascoigne, K. M. Proust, and M. O. Robins (Cambridge University Press 1990).

Nevertheless, by 1967, an intergovernmental AAT agreement was finalised. The design was based on a similar telescope then under development at the Kitt Peak National Observatory in Arizona. The 27.5-ton, 150-inch AAT mirror blank was cast at Owens-Illinois Inc. in Toledo, Ohio, from a low-expansion ceramic known commercially as Cer-Vit. The famous telescope manufacturer Grubb Parsons in Newcastle-upon-Tyne figured this mirror blank and designed and manufactured the main telescope tube. Mitsubishi Electric (Japan) provided the telescope mount, and the mechanical and electronic drive control. An emphasis on computer control, including an autofocus mechanism, was a pioneering aspect of the design. A technical project office was established in Canberra overseen by a Joint Policy Committee comprising a balance of senior UK and Australian astronomers. The one thing the agreement failed to mention was how the telescope would be operated. This was to be a source of disagreement for the next decade.

Fundamentally, the Brits and Australians disagreed on whether the AAT agreement specified only a telescope or, more generally, an *observatory* complete with an operating institution. After 1946, the Mount Stromlo Observatory had been placed under the aegis of the Australian National University in Canberra. Since Bart Bok had designated Siding Spring as the site for the AAT, many Australian astronomers argued it was unnecessary and wasteful of resources to establish an independent operating institution. In 1966 Olin Eggen, an American astronomer, became director of the (by now renamed) Mount Stromlo and Siding Spring Observatories (MSSSO) and was a forceful advocate for the view that overall management of the AAT, including the allocation of telescope time, should reside with him. However, the British were wary of losing control to the Australians, and Eggen himself was not overly popular in the Australian astronomical community outside of Canberra. Two British members of the Joint Policy Committee—Fred Hoyle, the famous Yorkshire-born astronomer then at Cambridge, and Margaret Burbidge, Woolley's successor as director of the Royal Greenwich Observatory—strongly countered Eggen's view. In a development that would surprise astronomers today, this UK-Australian dispute rose to the highest levels of government, with Margaret Thatcher, then secretary of state for education and science, pressing the British case against Eggen forcefully to her opposite number, Malcolm Fraser (later prime minister of Australia) in 1975.

Ultimately an intergovernmental reconciliation led in 1973 to the creation of a new institute, the Anglo-Australian Observatory (later the Australian Astronomical Observatory, AAO), with a director and staff distinct from MSSSO. However, the location of its headquarters then became the subject of a new dispute over whether to choose Canberra, Coonabarabran, or Sydney. The AAO was eventually established temporarily in Epping, within the Marsfield suburb of Sydney, a location finally made permanent in 1976.

Although modelled on the Kitt Peak telescope, the AAT was completed first (plate 15) and opened in 1974 in a ceremony presided over by Prince Charles. The prince was officially greeted by Fred Hoyle, whose vision for the AAT was to decisively shape the careers of many aspiring young astronomers, including myself. Up to that time, large telescopes were very much the province of experienced observers who had undertaken a decade or more of training on smaller telescopes. Appropriately for a theoretician, Hoyle believed that any professional astronomer, however inexperienced, should be able to apply for time to address a particular scientific question. Less experienced astronomers could be guided through the observing process by trained professional staff attached to the observatory. At the time, this was a profound change in the sociology of observational astronomy. It meant that a 28-year-old untenured postdoctoral researcher like myself, only 4 years past my PhD, could apply for time on one of the world's largest telescopes without going through years of apprenticeship at smaller facilities.

By the time I completed my PhD at Oxford in 1974, I was pretty disillusioned with academia. I didn't find studying the composition of the sun and nearby bright stars particularly inspiring. Much of this disappointment resulted from my own lack of confidence and the stultifying atmosphere in the Oxford Astrophysics Department at the time. I frequently reflect on that experience when advising my own graduate students today. PhD research is demanding and stressful, and the students engaged in it need constant encouragement, otherwise their morale will inevitably decline. Academic supervision in the 1970s was fairly haphazard by today's standards, and, in my case, I was more or less left to "get on with it." When I took a draft of my dissertation to my PhD thesis advisor for comments, he declined to read it, commenting that the document was entirely my responsibility.

For the first time since high school, I wavered in my ambition. Accordingly, I applied for several jobs outside academia. I focused initially on software companies, but my interviews did not go well. To this day I don't consider myself particularly good at writing computer code. However, I was gratified eventually to be offered a choice of two other positions. The first was with Macmillan, a famous London-based publisher responsible for the prestigious science journal *Nature*. However, in my case, I would oversee production of the magazine *British Birds*—hardly an obvious move after a PhD in astrophysics. The only thing they had in common is that both involved looking skyward. The second position was as an accounts executive in a highly regarded London advertising agency. This particular offer followed a gruelling series of four interviews, including one where I had to perform over lunch with a group of a dozen extroverted employees. Without question, the position would have been exciting and fast-paced, and the salary was "astronomical" by academic standards—an adjective that again was the only connection with the topic of my PhD! At the eleventh hour, however, my seemingly indifferent thesis supervisor drew my attention to a temporary teaching position in the Physics Department at Durham University, in a historic city in the north of England. The head of the Physics Department, Professor (later Sir) Arnold Wolfendale (1927–2020), was keen to move into astronomy. Although a cosmic-ray physicist, he eventually became Astronomer Royal. After a successful interview, Wolfendale's enthusiasm and surprising confidence in what I might accomplish at Durham convinced me to reject the other job offers and remain in academia.

After a challenging few months adapting to the Physics Department at Durham, where I was the only trained astronomer, I eventually embarked on a new research programme exploiting another telescope on Siding Spring mountain in Australia. Alongside the AAT is the UK Schmidt Telescope (UKST), installed in 1973 and, at the time, operated by the

Royal Observatory at Edinburgh (plate 16). Modelled on the same-sized Schmidt telescope at Mount Palomar, its primary task was to undertake a photographic survey of the southern sky. Although this survey involved taking and processing a uniform set of photographs by staff employed at the UKST, there was also the opportunity for individual visiting astronomers, including myself, to take additional photographs for their own research purposes. Each photograph taken with the UKST captured, in a single exposure, thousands of faint galaxies across an unprecedentedly large area of sky (equivalent to one that could contain nearly 130 images of the full moon). When I got my first look at one of these photographs, I was transfixed. Wherever I moved a magnifying glass over the photograph, there were numerous galaxies in addition to the occasional compact star. The brighter galaxies had spiral or elliptical forms, but there were also many smaller, presumably more distant, examples. At the limit of detection there were myriads of yet fainter smudges. It was the ultimate inspiration for me to want to explore the distant universe.

Traditionally, astronomers examined such glass photographic plates by illuminating them from behind and using a powerful eyepiece mounted on a moving carriage. However, this was not really appropriate for a Schmidt photograph, whose data content is enormous. In modern parlance, the Schmidt telescope is equivalent to a giant, 5000-megapixel imaging camera. In the 1970s, it simply wasn't practical to record all of its information digitally, and so "measuring machines" were developed to examine the photographs and process the imaging data into a more manageable catalogue of the positions, shapes, sizes, and brightnesses of every celestial image on the photograph. To do this, the negative photographic plate was rapidly scanned with a photomultiplier or laser, and the transmission of light through each pixel on the plate was analysed in real time. Pixels of varying darkness, due to the negative image of a galaxy or star, were processed to produce either a contour map of the darkness on the plate or a catalogue of sources and their shapes and sizes. These measuring machines were developed specifically to exploit what was sometimes called the "Schmidt telescope data revolution." In many ways it was similar to the current excitement over "big data" and "machine learning." In the United Kingdom, there were two rival measuring machines: the COSMOS machine at the Royal Observatory, Edinburgh, and the Automated Plate Measuring machine (APM) at the Institute of Astronomy, Cambridge.

I was very enthusiastic about this Schmidt revolution as the timing was perfect. The UKST was in its prime years of operation, and it was claimed that the quality and depth of its photographs were superior to those taken by the similar Schmidt telescope at Mount Palomar.[3] Moreover, the United Kingdom had the lead in the measuring-machine technology necessary to rapidly extract scientific results. In many ways, the UKST represented the birth of statistical extragalactic astronomy in Britain. Thousands of faint galaxies could be counted and studied over large swaths of sky to address key questions in cosmology and galaxy evolution.

With support from Wolfendale, a lively new group was formed at Durham to take on this research programme. It was headed by lecturer (assistant professor) Dick Fong, whose background in theoretical physics and mathematical ability were advantageous for the statistical aspects of this data-intensive project. During the latter part of the 1970s, with two PhD students, Steve Phillipps and Tom Shanks, we analysed both the counts and clustering patterns of faint galaxies on high-quality Schmidt plates scanned by the COSMOS machine at Edinburgh (plate 17).

Galaxies are clustered as a result of the gravitational forces acting between them. In a static, non-expanding universe, this clustering would inevitably grow stronger over time as there would be nothing to counter the attractive force of gravity. However, in an expanding universe, the rate at which the clustering grows is reduced, since there is a natural competition between gravity and the stretching of the space that moves galaxies further apart. Early work based on all-sky catalogues of nearby galaxies had established the extent to which galaxies are clustered. However, in the mid-1970s, a famous Princeton cosmologist, James "Jim" Peebles (Nobel Prize in Physics 2019), developed a mathematical theory for quantifying the clustering of galaxies in the context of theories of how cosmic structures grow with time. To test these theories would require

3. In this context, astronomers use the term "depth" to indicate the level of faintness or distance achieved in a survey, which, for the present story, is an approximate indicator of how far back in time is being explored.

measuring the clustering of much fainter galaxies seen at earlier cosmic times. This was a rather specialised discipline at the time, and even the Palomar Schmidt, which predated the UKST by over 20 years, had not embarked on serious observations of the clustering of faint galaxies. But for one example: in 1966 a group of Polish astronomers managed to convince Caltech to take some deep plates in an area of sky they proudly called the Jagellonian Field, after a dynasty of Polish kings. Without any measuring machines, three Polish astronomers painstakingly counted galaxies on these plates by eye, producing a catalogue of the positions of 15,650 faint galaxies. In 1977, together with his colleague Ed Groth, Peebles compared the clustering of galaxies in this Jagellonian catalogue, the deepest at the time, with that of brighter nearby galaxies.

Most of the Schmidt plates that our Durham team had analysed by this point were not taken specifically for our research and had one deficiency or another. Either the exposures were suboptimal, or the image quality was not perfect. As the feasibility and scientific promise of our project increased, it became clear we needed to secure the best possible plates specifically for our research. Accordingly, in 1978 I was sent to Australia to help with obtaining them. My host was Keith Tritton, officer in charge at the UKST. He and his wife Sue aided me in taking several excellent Schmidt photographs.

These UK Schmidt plates could reliably detect galaxies four times fainter than those in the Jagellonian catalogue, but even to the deeper limit achieved for five plates scanned by the COSMOS machine in Edinburgh, we were still unable to convincingly discover any change in the clustering of galaxies compared with that observed for nearby galaxies. In a comprehensive 1980 article led by Tom Shanks, we wrote that the strength of clustering is "in reasonable agreement with those expected from . . . local results, indicating the uniformity and isotropy of the Universe to very great depths. However, the errors are still too large to put useful constraints on evolution in galaxy clustering."[4]

---

4. Correlation Analyses of Deep Galaxy Samples. II. Wide Angle Surveys at the South Galactic Pole, T. Shanks, R. Fong, R. S. Ellis, and H. T. MacGillivray, *Monthly Notices of the Royal Astronomical Society*, vol 192 (1980), p209.

After four years of group effort on this clustering project, it seemed there were too many unknowns. All we could measure on these Schmidt plates were galaxy positions and apparent brightnesses (figure 4.2, *left*); we had no distance (or redshift) information, so there was no direct indication of the lookback times involved. How, then, could we quantify over what time interval any growth of clustering had occurred? Moreover, by looking back in time, there was a big unknown. Conceivably, the luminosities of galaxies might be different, which would undoubtedly affect our estimates of the volumes involved and hence any interpretation of clustering evolution. Nonetheless, our group had learned much about the construction and interpretation of catalogues of faint galaxies, and our articles received a lot of attention—and not just in the United Kingdom, where we were already a growing influence in observational cosmology. During a bitterly cold January 1978, I toured the East Coast of the United States advertising the work of the newly founded Durham group. That's when I met Jim Peebles for the first time, and he showed great interest in our work.

During a visit to Cambridge later that year, I met Bruce Peterson, an American astronomer then working at the AAO. He had a high-quality photographic plate taken at the prime focus of the 3.9-metre AAT. As the AAT is a much more powerful telescope than the 1.2-metre UKST, this photograph reached to much fainter galaxies, but over a smaller area of sky. We processed the information on this plate using software developed by the Cambridge group that operated the APM machine and analysed the counts of galaxies to fainter limits than hitherto (figure 4.1). Although conscious we were following in the footsteps of Hubble (chapter 3), it was generally accepted by now that galaxy counts were useful probes of galaxy evolution rather than the curvature of space. With the advent of improved photographic emulsions developed by Kodak and using clever tricks to hypersensitise them developed by the AAT's talented photographer David Malin, our galaxy counts reached a depth 100 times fainter than Hubble's in 1936.

To test for evolution in the faint galaxy counts, we had to construct a "no-evolution" prediction. In the jargon of a statistical experiment, this is what is called a "null hypothesis." Given the faintness limit of our

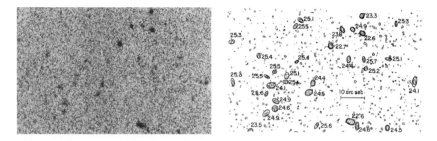

FIGURE 4.1. (*Left*) A small portion of the (negative) photographic plate taken by Bruce Peterson at the AAT. (*Right*) A contour map resulting from the processing of this area by the Cambridge APM group. Numbers refer to apparent magnitudes (a logarithmic indicator of brightness with increasing values meaning fainter sources).

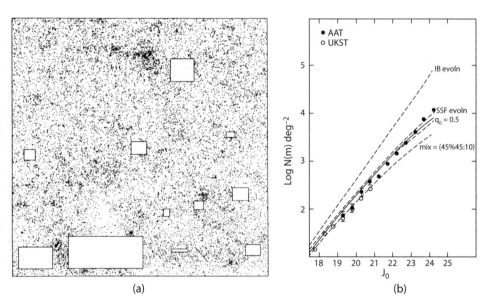

FIGURE 4.2. Measuring-machine studies of photographic plates in the 1970s. (*Left*) The sky distribution of thousands of faint galaxies on a single Schmidt plate scanned by the COSMOS machine. Blank rectangular areas represent portions affected by very bright stars or artificial satellite trails. This was used to analyse the clustering of galaxies as a function of limiting brightness. (*Right*) Galaxy counts from Bruce Peterson's AAT plate (solid circles) compared with those from several Schmidt plates (open circles). The *y* axis is the logarithm of the number of galaxies per unit sky area, and the *x* axis is the astronomical magnitude—a logarithmic measure of faintness (increasing to the right). The IB and SSF curves represent Tinsley's "instantaneous burst" and "slow star formation" models. The other curves represent various no-evolution predictions. The data suggested early evidence for evolution in the galaxy population.

catalogue, we predicted, for a non-evolving galaxy of a given luminosity, how far away it could be seen and therefore the relevant volume that the photograph would encompass for sources of that luminosity. Of course, not all galaxies have the same luminosity. However, from local galaxy surveys, we knew the volume density of galaxies of different luminosities, a property called the "luminosity function." Naturally, the most luminous galaxies could be seen to the greatest distances and hence over the largest volumes, but they are rare in terms of their number per unit volume. Feeble galaxies are more numerous per unit volume, but they can't be detected if they are far away, even in deep photographs. By adding up the numbers in the various volumes across the full range of the luminosity function, we could predict, for various faintness limits, how many galaxies we should expect to count on our photographic plate in the absence of any luminosity evolution.

Remarkably, we found more faint galaxies than expected on the basis of this no-evolution prediction (figure 4.2, *right*). As we counted progressively fainter galaxies, their numbers increased *faster* than would be the case in a non-evolving universe. At first glance this may seem counter-intuitive. If galaxies were still assembling, surely there would be fewer in the past? However, if the earlier galaxies were more luminous, we would be able to see further into the past and thus would *underestimate* the volume reached for a given faintness limit in the no-evolution calculation. In an article published in the *Astrophysical Journal* in 1979, we concluded that the data "indicated a moderate amount of luminosity evolution."[5]

At the same time, Beatrice Tinsley at Yale University (chapter 3) was attempting to model and predict such results. Since stars are born, evolve, and die, and the optical light we see from a galaxy is the sum total of all its stars, it follows that the galaxy's luminosity will be closely related to the rate at which it forms new stars. If, for example, all the stars in a galaxy formed in one instantaneous burst of activity a long time ago, the stellar population would age uniformly. Young luminous stars are

5. Number Magnitude Counts of Faint Galaxies, B. A. Peterson, R. S. Ellis, E. J. Kibblewhite, M. T. Bridgeland, T. Hooley, and D. Horne, *Astrophysics Journal Letters*, vol 233 (1979), L109.

short-lived because they burn their nuclear fuel quickly and rapidly explode as supernovae. Thus, as time progresses, only the dimmer, long-lived, stars would remain. In this simple case, a galaxy would be spectacularly luminous at first, but fade rapidly thereafter. We argued such a dramatic model of galaxy formation overpredicted our galaxy counts (figure 4.2, *right*). On the other hand, if stars formed slowly over time, rather than all at once, the stellar population would be constantly replenished with newly born, young, luminous stars and the resulting galaxy would decline in its luminosity more gradually. This was a better match to our data. Although there were a variety of possible interpretations, by finding *an excess number of faint galaxies* we had conclusive evidence that at least *some* galaxies were more luminous or more numerous in the past.

In comparison with observational astronomy at the frontiers today, the scene was nowhere near as competitive or stressful. There were only two other groups in the game of faint-galaxy counting, and we soon got to know one another. At University of California, Berkeley, counting galaxies on photographic plates was led by a modest but intellectually sharp graduate student, Richard Kron. His thesis on the topic was so impressive he immediately landed a professorship at the University of Chicago. There was a second group at Bell Laboratories, the famous research institution in New Jersey originally funded by the American telecommunications giant. This was led by Tony Tyson, a creative individual with an entrepreneurial flair. Scientific articles by Tyson and Kron appeared soon after our article. Tyson's results agreed with ours, but Kron adopted a different approach to measuring the brightness of a faint galaxy, so it was hard to compare his results.

As the night sky is not completely dark, it is not straightforward to measure the total light of a faint galaxy. Considering a contour map such as in figure 4.1 (*right*), it is possible to detect a faint galaxy only down to a certain surface brightness level (termed "isophote"), rather like measuring the volume of a mountain down to some intermediate elevation rather than the ideal value down to its base or sea level. We attempted to address this "missing-light" effect by comparing our observations with models that predicted the brightness above this (known) isophote.

In contrast, Kron extrapolated the light profile of each galaxy below what could be readily detected in order to estimate the total brightness. This led to many philosophical arguments between Kron and myself as to whether it was better to model imperfect observations or correct actual data to its ideal equivalent.

The published galaxy counts from the three groups received much attention, as this was one of the first pieces of direct evidence that galaxies were indeed evolving as Tinsley had predicted. Following the failed Palomar campaigns to use galaxies to probe the expansion history of the universe (chapter 3), the frontier had now clearly moved to studies of how galaxies formed and evolved. A particularly memorable conference was held at the University of California at Los Angeles in the summer of 1979, entitled "Objects of High Redshift." The organisers were Jim Peebles and George Abell, the latter a distinguished observer who had used the Palomar Schmidt sky survey to make a comprehensive catalogue of distant clusters of galaxies. On the opening morning of the conference, at the youthful age of 29, I gave the results of the Durham group's work on the evolution of galaxy clustering as well as the implications of the counts on Bruce Peterson's plate. The audience was packed with famous astronomers, so this was personally very inspiring. The conference dinner was a barbecue on Mount Wilson, where we saw the famous 100-inch where Hubble observed, and there was an excursion to Disneyland where I just about survived the roller-coaster ride at Space Mountain. This was my first visit to California, and I was on an all-time high.

While it was reassuring to be invited to international conferences and present the Durham group's results, I was trying to think ahead and ask, "what next?" Reflecting that all three galaxy-count groups were claiming evidence for some form of evolution, it seemed clear to me that we had reached the same impasse as with the earlier clustering of galaxies project. We had no direct information on the lookback times being probed, and thus no end of modelling by Tinsley or ourselves could uniquely determine *how* galaxies were evolving, or even, fundamentally, *how far back in time* these photographic plates were probing. The only resolution of this dilemma would be to try to measure the redshifts of the galaxies being counted.

During these early AAT years, there was a healthy race with US telescopes in the spectroscopy of ever-more-distant quasars. Although the AAT's 150-inch mirror was smaller than that of the Hale 200-inch, the AAT had a remarkable detector called the Image Photon Counting System (IPCS), developed by Alec Boksenberg at UCL, which gave it a cutting edge. I knew Boksenberg very well as he had been one of my undergraduate lecturers. Sociable and with a puckish sense of humour, he was a wizard with electronics. The IPCS was a major advance over the photographic plate in sensitivity and had a particularly interesting and unique feature. Unlike most detectors, where one initiates a long exposure and waits patiently for it to finish and display the result, the IPCS counted photons and displayed them in real time. The astronomer could thus examine the signal as the data were coming in. Not only was observing with the IPCS more exciting, but an astronomer could adjust the exposure time while assessing the incoming signal. In the early 1980s, the AAT, together with the survey capability of the UK Schmidt, was rivalling Californian dominance in spectroscopic studies of the most distant quasars, detected by now up to record-breaking redshifts beyond $z \sim 4$. I was hopeful we could do the same using the IPCS detector to gather the redshifts of large numbers of the faint galaxies seen on our deep photographic plates, and thereby break the impasse in understanding the origin of these excess faint galaxies.

On scientific projects unrelated to faint galaxies, I was becoming a frequent observer at the AAT. Given the absence of a fast internet in the 1980s, it was essential to visit Siding Spring and observe in person. In the late 1970s and early 1980s, it was a 24-hour flight from London to Sydney, with brief intermediate stops at Bahrain and Singapore. Having flown as a student only in Europe, I remember my first trip "down under" as exciting, but also an ordeal. However, after I started doing such long trips from London three or four times a year, it became routine. I became an expert at avoiding check-in baggage, selecting the right seats, and keeping busy during long-haul flights. This was a time when many British astronomers were flying over 100,000 air miles a year (and before the advent of frequent-flyer clubs). One British astronomer bragged he'd visited the AAT "for the weekend"! Alec Boksenberg, whose IPCS was eventually installed on both the AAT and Palomar 200-inch, proudly claimed *he'd observed on both telescopes on the same night*, crossing the International Date Line to accomplish the feat.[6]

Sydney, Australia, was once claimed to be the best address in the world, and indeed it is a fantastic city with amazing restaurants, sparkling beaches, and a glorious climate. Despite the 10,500-mile air trip, it seemed strangely familiar on arrival: pharmacies were called "chemists," driving was on the left, and Queen Elizabeth's portrait was on the coins. An astronomer at the AAO once told me that Australia is "California run by the Brits." However, Siding Spring Observatory is not in Sydney, it is in what Australians call "the outback." To get to the AAT involved a 1-hour domestic flight from Sydney to Coonabarabran, often via a bumpy ride on a propeller plane. The pilot always did a first pass over the runway to warn off kangaroos. Driving through "Coona," a pleasant country town of just over 2000 residents, was like going back two to three decades. The contrast with the buzz of Sydney was striking. The taxi driver who had the contract to take observers to the telescope, a ride of 40 minutes, was called Charlie. There was an obligatory stop for beer and wine at a liquor store before driving into the beautiful Warrumbungle National Park, taking care to avoid iguanas sunbathing in the road. Charlie appeared to everyone as the authentic "ocker."[7] Diminutive in stature, his oversized beer gut protruded over the driving wheel, beneath which he sported brown "stubbies" (short pants) and suntanned legs. Charlie entered the folklore of AAT observers. I like to believe he deliberately aimed to shock visiting, delicately reared, rosy-cheeked English students with his raw Aussie banter. After travelling with Charlie maybe 100 times over the years, we got to understand each

6. On closer examination, Boksenberg would have been capable only of observing on the AAT and Palomar on the same day. He would leave the AAT at dawn, fly to Sydney, and catch an afternoon direct flight to Los Angeles, arriving on the morning of the same day. A car ride to Palomar would get him there in time for sunset. This is not strictly the same night, but a miraculous feat nonetheless. Whether he was in a fit state to observe after such a heroic trip I didn't enquire!

7. Defined in the Oxford Dictionaries as "a rough, uncultivated Australian man" (Oxford University Press, https://premium.oxforddictionaries.com/definition/english/ocker [accessed via Oxford Dictionaries Online on January 26, 2022]).

other pretty well. He was always there on time with a cheerful grin and his latest outrageous commentary on life in the outback.

My early observing runs at the AAT were inspirational. There really was a sense of exploration and excitement when using a first-rate new telescope with the IPCS detector. Much of my early observing was done with George Efstathiou, an Oxford undergraduate who, by our good fortune, had decided to come to Durham as a PhD student. George was that rare individual, even as a student, who had the mathematical and physical insight to become a first-rate theorist but could appreciate the importance (and joy) of going to the telescope. The AAT moved precisely and effortlessly from one target to the next, so accurate positions were essential. The telescope operators were friendly and positive, and there was a supportive atmosphere (plate 18). From the lofty catwalk, high above the eucalyptus forest below, we could peer far westward into the nothingness of central Australia. Witnessing me contemplating this view at sunset, an AAT engineer once commented: "Ah, staring at the Great Australian Bugger All!" Following a successful night of observations we would retire for a beer or two near the central canteen, keeping clear of grazing kangaroos on the way since they had been known to injure less careful astronomers. There we would compare notes with astronomers using the other telescopes on Siding Spring, before trying to sleep until noon. After a productive two or three nights on the telescope, we'd board the plane back to Sydney loaded with a box of magnetic tapes containing raw data. During a week at the AAO headquarters in a suburb of Sydney, we would process this raw data into scientific results. Often we were invited to the homes of AAO staff for an evening meal. Despite the effort of processing our hard-won data, there was still ample time to enjoy the fine weather, maybe visit a beach, and buy souvenirs for my two small children. We'd return to Durham with enthusiastic stories of our scientific discoveries and adventures down under.

Returning to my plans to embark on a faint galaxy redshift survey, the challenges were formidable. Whereas the AAT was keeping up with those competitive Californians in studies of quasars, a luminous quasar at a redshift $z = 4$ is more than 250 times brighter than the faintest galaxies on Bruce Peterson's photograph. Even though many of those galaxies were probably at a lower redshift, securing their spectra would be much harder. To construct a representative redshift survey of over 100 faint galaxies would take dozens of AAT nights, whereas my typical observing allocation was only three or four nights a year. At about this time, I attended a talk at the Royal Astronomical Society given by Roger Angel, a British astronomer at the University of Arizona. He described a new technique based on using fibre-optic cables to gather the light from several galaxies in the field of view of a large telescope and assemble these fibres along the entrance slit of a spectrograph. In this way Angel and colleagues had secured the spectra of many galaxies simultaneously. I was inspired by Angel's talk and, together with a collaborator, David Carter, who had just moved to the AAO, in 1981 wrote to the director to ask if such a facility could be provided at the AAT. The director, Donald Morton, was always receptive to new ideas, particularly from junior astronomers. He instructed a young Australian engineer, Peter Gray, to work with us on implementing Angel's idea. Remarkably, within only a few months, Peter had built a makeshift system feeding 21 fibres from holes drilled into a brass plate mounted at an auxiliary telescope focus to the entrance slit of an AAT spectrograph (plate 19, *left*). With only a 1.5-hour exposure, we secured the redshifts for all 21 target galaxies. This would have taken three nights without this multiplex advantage. Other astronomers took note of our success and demanded access to this novel capability for their own research. The simultaneous observations of 50 targets became routinely possible soon thereafter.

However, operation of this multi-object spectroscopical capability was a little cumbersome. Astronomers had to deliver precise sky positions for their targets to the observatory staff, who would use a milling machine to drill holes accordingly in a brass plate. On the night, astronomers would manually plug 50 fibre-optic cables into this plate and insert the plate into the telescope focal station. Each fibre had an identification number corresponding to the order in which it was aligned at the entrance slit of the spectrograph. Furthermore, each hole drilled in the brass plate had a different number, corresponding to the position of a particular celestial target. The association between fibre number and target number had to be meticulously recorded on a sheet of paper so the astronomer knew which spectrum on the detector came from which astronomical target (plate 19, *right*). Woe betide you if you lost that important piece of paper!

With the success of this fibre-optic system, in 1985 Tom Shanks and I embarked on our first faint galaxy redshift survey with Tom Broadhurst, a graduate student. The goal was to determine the extent of the cosmic volume containing the excess galaxies seen in the counts on Bruce Peterson's AAT photographic plate. By now several AAT photographs had been taken, further confirming the excess numbers originally seen with Bruce Peterson's plate. Plates had been taken separately through blue and red filters, and the excess was more pronounced on the blue photographs. The puzzle of the count excess became known as the "faint blue galaxy problem." Shanks and I realised it was too ambitious to attempt spectroscopy of the very faintest galaxies seen on these AAT plates, so we initially aimed to complete a spectroscopy survey at the limiting depth of the UK Schmidt plates where the number count excess first became apparent. Over several observing runs, Tom Broadhurst and I secured over 200 redshifts in five different areas of sky, with typical exposure times of 6 hours. Tom Broadhurst was a great companion in Australia, and the survey formed the basis of his PhD thesis (plate 20, *left*).

What did we find from this survey? If galaxies were more luminous in the past, as would be the case if they were forming stars more vigorously in their youth, our survey would reach to higher redshifts than predicted in the absence of evolution. Alternatively, Richard Kron had hypothesised that the excess counts might reflect a poor understanding of the numbers of feeble, or "dwarf," galaxies nearby. In this case, our survey would reveal that many of the faint galaxies on our photographic plates were actually at low redshifts. Neither explanation matched our data. Our survey revealed a distribution that extended to a redshift $z = 0.45$. This indicated we were probing to a lookback time of about 4.8 billion years (the past 35% of cosmic history). There was no surprising population of nearby galaxies, so we believed we could exclude Kron's hypothesis that the excess numbers of faint galaxies arose from previously undetected nearby dwarf galaxies, but neither did our data provide evidence for strong luminosity evolution for the overall population. Basically, the redshift distribution was consistent with the no-evolution prediction (figure 4.3, *top*).

The outcome was quite puzzling. Although there was clearly an excess population of galaxies seen on the photographs, the cosmic volume being probed by our redshift survey was as expected if there were no excess. It was as if some "extra" galaxies present 5 billion years ago had somehow disappeared by the present day. A review of our work in the influential journal *Nature* was entitled "Galaxies: Then You Saw Them, Now You Don't."[8]

The scientific article discussing our results, led by Tom Broadhurst, bravely suggested that perhaps only *some* of the galaxies were brighter in the past. To support this conjecture, we noted that our spectroscopic data revealed that many galaxies displayed strong gaseous emission lines consistent with a recent burst of star formation. If this burst-like activity was more prevalent in the past, it would temporarily increase the luminosities for a subset of our sample. We argued that only the *less luminous galaxies* were suffering these intermittent bursts. Many theoreticians were sceptical of this complicated picture, but, as observational astronomers, we had concocted the only empirical model consistent with our data. However, those modelling galaxy evolution wanted a "neat-and-tidy" universe where all galaxies evolved in a similar manner. The notion that some "rogue" subset of the population was capable of doing its own thing seemed contrived.

As more astronomers recognised the huge gain of multi-object spectroscopy, those of us at Durham began to consider how to automate the positioning of fibres in the focal plane of the telescope. Not only would this avoid the need for drilling holes in brass plates and the tedious paper accounting mentioned earlier, it would also prevent damage to the fragile optical fibres that frequently occurred from their repeated plugging and unplugging from numerous brass plates. It was at this time I realised the crucial advantage of investing effort in innovative instrumentation. A creative fellow Welshman in our fledgling Durham

8. R. S. Ellis and C. S. Frenk, *Nature*, vol 346 (1990), pp790–91.

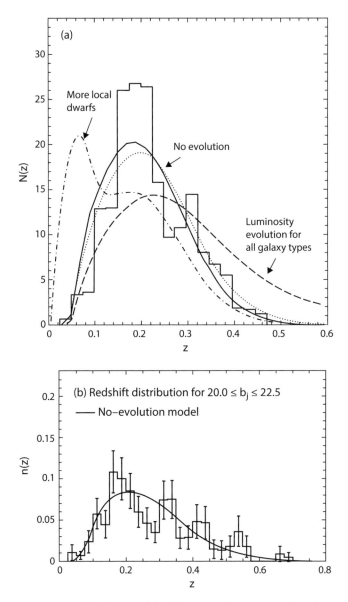

FIGURE 4.3. The redshift distribution $n(z)$ of the AAT galaxy surveys indicated no significant deviation from a no-evolution distribution, despite a substantial excess in the number counts. (*Top*) The Autofib survey excluded the hypotheses that the excess counts represented an unforeseen population of local dwarf galaxies, or luminosity evolution for all galaxies. (*Bottom*) The LDSS survey extended the campaign to fainter sources where the count excess was much larger. The combined sample of 350 faint galaxies indicated a redshift distribution that still matched the no-evolution prediction. In order to explain the excess counts it was proposed that only a subset of feeble star-forming galaxies were more luminous in the recent past. Galaxies were evolving at different rates according to their masses.

instrument team, Ian Parry, began to consider how to design an auto-mated fibre positioner.

Roger Angel's group at the University of Arizona were also develop-ing an automated system, called the MX spectrometer. The latter instru-ment had 32 robot arms, arranged around a circumference, that could simultaneously place 32 fibres within a circular field. In referring to their design, they introduced the analogue concept of "fishermen around a pond." Although MX's simultaneous positioning saved precious tele-scope time, it turned out to be geometrically difficult to allocate equally spaced robot arms to randomly distributed targets inside the field. The arms could move radially from the circumference into the centre but had limited lateral motion. Moreover, the arms could not cross one an-other. Celestial targets do not always arrange themselves on the sky in a convenient pattern for such an instrument. Finally, a multiplex gain of 32 did not represent any progress over the capability we already enjoyed with the brass plug plates at the AAT.

Ian's alternative design was inspired, even if its name, Autofib, was not. He adopted a "pick-and-place" approach, whereby an electromag-net on a moving x-y carriage picked up and deposited a small "button" attached to the end of each fibre cable. This button contained a tiny rare-earth magnet (which firmly held the button to a steel plate when deposited) and a right-angle prism, which reflected light from the sky along the fibre (plate 21). Autofib was initially built with 64 fibres (it was later extended to 150), which could cross one another if necessary, giving valuable flexibility in assigning fibres to targets. The only minor drawback compared to the MX design was that it took 8–10 minutes to configure the fibre buttons, one by one, onto 64 targets. This was a small price to pay given the other advantages.

Although our Durham team continued further redshift surveys with Autofib, these fibre-based campaigns could not push to fainter galaxies because of limitations arising from the ubiquitous signal coming from the night sky. For the fibre-based observations undertaken with Broad-hurst and Shanks, the night-sky signal was determined by averaging the spectrum obtained with a set of fibres dedicated to blank sky; this was then subtracted from each galaxy spectrum. At the faintness limit of that survey, the sky signal was comparable to that arriving from a galaxy. Unfortunately, however, optical fibres do not transmit with uniform efficiency, and although we tried to correct for these variations by carefully measuring their throughputs with calibration lamps in the afternoon, such sky subtraction challenges indicated we could not probe much fainter with Autofib.

A traditional spectrograph has a rectangular entrance aperture in the shape of a long slit. The sky signal adjacent to a galaxy can be averaged along this entrance slit, ensuring accurate subtraction. However, the spectrographs in common use in the mid-1980s were capable of studying only one galaxy at a time, or occasionally a close pair when the slit was aligned to target both sources. Clearly, what was needed for furthering our studies was a *multi-slit* spectrograph. This could be accomplished if, instead of having a single entrance slit as in a traditional spectrograph, a metal plate cut with many individual short "slitlets" were placed at the telescope focus. The light passing through each slitlet would produce its own spectrum of the galaxy so targeted, as well as provide a measure of the adjacent sky signal. Ideally, the multiplex gain should be similar to that which had been revolutionised through the use of fibre optics. Such an instrument would require a large field of view and thus be very challenging to design, but it would enable improved sky subtraction and help us reach fainter targets efficiently.

From 1983 to 1985, I took leave of absence from my lectureship (assistant professorship) at Durham to take up a position at the Royal Greenwich Observatory (RGO) in Herstmonceux, Sussex. The recently appointed director was Alec Boksenberg, inventor of the IPCS and now something of a national hero. The RGO was charged with developing the newly established observatory on La Palma in the Canary Isles, which I discuss in more detail in chapter 5. At the RGO I met Keith Taylor, a talented astronomer who, with a colleague at Imperial College, Paul Atherton, had pioneered an instrument called TAURUS. TAURUS was a clever imaging system that, using a scanning system called a Fabry-Perot interferometer, could map the spatial and velocity structure of gaseous emission across nearby galaxies. Given Keith's experience with this imaging system, we focused our attention on how to design an equivalent wide-field instrument for faint multi-slit spectroscopy. This became LDSS—the Low Dispersion Survey Spectrograph.

It took a while to adapt to working at the RGO. As it was part of the scientific civil service there was much bureaucracy, and, given its largely technical purpose, only a small fraction of staff had the freedom to conduct research. This was a big contrast with Durham, where the corridors of the physics department were often full of noisy, long-haired undergraduates and there was a genuinely youthful atmosphere. Many RGO employees were skilled engineers or computer programmers with clear, well-defined tasks. They worked from 9 a.m. to 5 p.m. and made sure they took all the holidays (and perks) that came with working for the scientific civil service. In fact, I got into trouble for not taking enough holiday, as it made an administrator's accounts impossible to reconcile! I got to like Keith Taylor because, somehow, he managed to effectively straddle these two very different worlds. He was fully conversant with how RGO worked, yet had a wild, creative streak consistent with being an academic researcher.

The RGO employed two talented optical designers, Professor Charles Gorrie Wynne (1911–1999) and his assistant, Sue Worswick. Wynne had a formidable reputation; he was one of the first to apply computational methods to optimising the design of a variety of optical systems. For example, it was Wynne who pioneered a three-lens glass corrector that enabled the AAT to deliver high-quality photographic plates such as the one I had analysed with Bruce Peterson. On the other hand, like many highly talented individuals, Wynne could be stubborn and uncooperative. Taylor asked Wynne and Worswick to design a spectrograph with the maximum possible field of view at the Cassegrain f/8 focus of the AAT. This would yield the largest multiplex gain and be a very effective survey instrument. Keith certainly knew how to charm and get the best out of Charles Wynne, once commenting that "optical designers need constant grooming like racehorses." The resulting design did not disappoint. LDSS would have an elephantine field of view of 12 arcminutes across, unprecedented for such an instrument on a large telescope. The breakthrough was the use of a new type of glass called FK54 from the famous German manufacturer Schott. Although its low-dispersion chromatic properties were exceptional, FK54 was a delicate fluorophosphate glass with a high thermal expansion coefficient. Imperial College Optical Systems, the company chosen to figure the material into the lenses for LDSS, encountered much difficulty shaping this glass for our spectrograph.

Once again, we had very few competitors in the race for multi-slit spectroscopy of faint galaxies. The European Southern Observatory (ESO) pioneered a similar spectrograph called the ESO Faint Object Spectrograph and Camera (EFOSC) on their 3.6-metre telescope at La Silla in Chile. Commissioned in 1984, it was the first spectrograph to exploit the novel properties of Schott's FK54 glassware, but had a much smaller field of view. EFOSC was designed for the general community and so offered a variety of imaging and spectroscopic options, whereas LDSS was a basic, "no-frills" instrument purposely designed for our galaxy survey. At Kitt Peak, Richard Kron and another Berkeley student, David Koo, were undertaking a similar survey using an instrument called the Cryogenic Camera, which also had a more modest field of view. EFOSC and the Cryogenic Camera used a charge-coupled device (CCD) as a detector, whereas LDSS was capable of using either a CCD or Boksenberg's IPCS. Although the IPCS was less efficient, it had a larger format and offered real-time data inspection. LDSS was commissioned on the AAT in 1986, by which time I had returned to a full professorship at Durham.

The entire LDSS project was accomplished on a shoestring. Its construction was funded by a UK research council grant of only £18,400 ($27,000 at the time). All aspects of the instrument were manually controlled, which meant observers had to ride in the AAT Cassegrain cage throughout the night. The field of view was so large that acquiring faint galaxies precisely onto their individual slits necessitated dedicated software, both in carefully designing the multi-slit masks and for the operation at the telescope (plate 22, *left*). With great foresight, Keith realised this challenge and brought a talented programmer, Richard Hook, onto the team (plate 20, *right*). During a night's observing with LDSS, we took turns to lie on a mattress in the darkened Cassegrain cage for hours at a time, getting instructions by walkie-talkie from the control room. As the long exposures progressed the telescope tilted to track our targets, and we had to shift the mattress accordingly, often into an uncomfortable, almost vertical, position (plate 22, *right*).

The LDSS redshift survey data were analysed by a very smart and astonishingly well- organised postdoctoral assistant, Matthew Colless. Matthew hailed from Grafton in New South Wales but garnered his PhD in Cambridge. (After working with me in Durham, he returned to Australia and rapidly rose up the ranks, becoming the director of AAO and subsequently that of the Australian National University Research School of Astronomy and Astrophysics, formerly MSSSO). Together at Durham we secured redshifts for 150 galaxies, penetrating nearly three times fainter than we had with the fibre-optic system. Our new survey reached galaxies with redshifts as high as $z = 0.7$ (a lookback time of over 6 billion years, or nearly half the present cosmic age). The redshift distribution continued the puzzling no-evolution trend consistent with the results we'd published with Tom Broadhurst. At the deeper faintness limit reached by LDSS, the galaxy counts were now fully twice those predicted in a no-evolution case, yet, once again, the redshift distribution was consistent with no evolution (figure 4.3, *bottom*). If galaxies of all luminosities were brighter in the past, we would have detected galaxies far beyond a redshift $z = 0.7$. We thus could confidently exclude models where all galaxies evolve monotonically with redshift in a luminosity-independent way. Our preferred explanation remained that luminosity evolution was restricted to a sub-luminous component whose enhanced star formation was evident from intense gaseous emission lines. These sources, we argued, represented the excess numbers in the galaxy counts.

Our results implied that the most massive, luminous galaxies were largely unchanged over the past half of cosmic time. Instead, most of the evolution was occurring in feeble galaxies that were forming stars at a prodigious rate only 6 billion years ago. By 1990, when we published our results, most theorists imagined that galaxies assembled hierarchically from smaller systems drawn together by their gravitational attraction. Our picture suggested the opposite. The massive galaxies were largely assembled before $z = 0.7$, whereas the less massive galaxies were still actively forming their stars in the recent past. This progressive "downsizing" of activity from large galaxies to lesser ones later became the standard picture. It demonstrated that galaxy assembly was not governed just by gravity but involved additional processes that determined the rate at which gas clouds condensed to form stars.

Without question, multi-object spectroscopy became a major highlight of AAT science in the late 1980s and 1990s. Autofib and its plug-plate predecessor were used in numerous other high-impact programmes. These ranged from a landmark survey of over 400 faint quasars, which transformed our understanding of how those systems evolved from redshifts of $z \sim 2$ to the present day, to the kinematics of hundreds of stars, which demonstrated that the dark matter in the Milky Way must reside in a large extensive halo rather than in its galactic disc.

However, the technical *pièce de résistance* was yet to come. From 1985 to 1986, the Royal Astronomical Society undertook a forward-looking survey to consider what facilities the UK community needed in the 1990s and beyond. As a recently appointed full professor at Durham, I was invited to serve on this committee. It was chaired by Sir Francis Graham-Smith, a distinguished radio astronomer and, at the time, the Astronomer Royal. Another member was Donald Lynden-Bell (1935–2018), a formidable theorist at Cambridge who was promoting a dedicated wide-field spectroscopic survey telescope that his colleague Roderick Willstrop had designed. Willstrop was actually one of the commissioning astronomers at the AAT in the early 1970s and a talented optical designer.

Willstrop's proposed telescope had a field of view on the sky almost 50 times that of the AAT. Following the publication of the RAS report, I was asked to chair a committee to consider various options for such a wide-field telescope. We consulted several engineers and optical designers, including Wynne, Willstrop, and Parry, and in 1987 proposed that it would be more economical to upgrade the optics of the AAT rather than build a completely new telescope. The proposal would increase the AAT's field of view fourfold via a new glass corrector at its prime focus. We also recommended an enhanced version of Autofib comprising 400 optical fibres feeding two new spectrographs, a significant gain in capability. As the large, new glass corrector would have a field of view two degrees in diameter (large enough to encompass 16 full moons), the project became known as the 2-degree field project, or 2dF.

To promote this ambitious 2dF facility, I had to present the scientific case to the AAT board of directors in 1988. I did this with gusto, after the board had enjoyed a heavy lunch with several bottles of wine. Although a few members fell asleep, support was nonetheless forthcoming! A design study conducted by Keith Taylor (who moved to the AAO purposely to realise this facility) and Peter Gray was approved in 1989, subject to raising A$1.6 million (later A$2.1 million) equally from UK and Australian funding agencies. The instrument was commissioned in 1997 (plate 23). Matthew Colless (by then at MSSSO) and I led a UK-Australian science team that conducted a survey of 250,000 bright galaxies, largely for cosmological purposes, during 1998–2003 (plate 24). When I emigrated to the United States in 1999, I relinquished the UK co-lead to John Peacock, a cosmologist at Edinburgh.

Although this 2dF Galaxy Redshift Survey did not probe to significant lookback times and therefore does not constitute a major part of the present story, it is nonetheless probably the most significant achievement of the AAT in extragalactic astronomy. In addition to a definitive measurement of the dark matter content of the universe, coincident with a similar US survey (the Sloan Digital Sky Survey), the 2dF survey located a feature in the galaxy distribution on very large scales that is the relic of one imprinted in the early universe. Predicted to exist by theoretical cosmologists and now confirmed observationally, this feature acts as a "cosmic ruler" that grows in size with the expansion of the universe. This large-scale feature is now being traced by later survey instruments over a wide range of redshifts, and hence lookback times, to provide the most precise measurements of the history of the cosmic expansion. The prestigious 2014 Shaw Prize in Cosmology was jointly awarded to those in the 2dF and Sloan Digital Sky Survey teams who pioneered these initial cosmological measurements. In total the 2dF team published 42 scientific articles, which have been collectively cited by other scientists over 13,000 times. When the survey data were made public it led to over 270 further papers by external scientists.

From 1978 to 1999, prior to my departure to California, I clocked up over 150 observing nights on the AAT and undertook more than 70 visits to Australia. Not all trips were concerned with observing because,

from 1991 to 1995, I also served as a member of the AAT board of directors. Like many UK astronomers of my generation, this telescope
launched my professional career in observational astronomy. It also
taught me the importance of developing innovative instrumentation
motivated by key scientific questions. Through the partnership with
Peter Gray and Ian Parry, my colleagues and I were able to use multi-
fibre technology to commence our faint galaxy redshift surveys; others
used it to make similar astronomical advances. Later, through the ingenuity of Charles Wynne and the enthusiasm of Keith Taylor, we used
LDSS to push studies of evolution halfway back to the Big Bang. It is no
exaggeration to say that the AAT became the envy of both European
and American astronomers. Not only was the telescope at the forefront
in the 1990s but, through the 2dF project, it was a pioneer in demonstrating the benefits of rejuvenating a 25-year-old facility. Such rejuvena-
tion of 4-metre-aperture telescopes with new instrumentation later
became commonplace elsewhere as a new generation of 8- to 10-metre
telescopes arrived in the mid- to late 1990s.

5

La Palma

La Isla Bonita

The Observatorio de Roque de Los Muchachos (the name rolls off your tongue after you've observed there a few times) is in a truly magnificent location. Perched near the summit of La Palma, the northwesternmost island of the Canary Islands, it is operated by the Instituto de Astrofísica de Canarias (IAC) based at La Laguna on the larger island of Tenerife. The Spanish name refers to the "Rock of the boys," an outcrop at the island's summit at an elevation of 2400 metres. The observatory is precipitously close to the edge of the Caldera de Taburiente, a massive crater 1500 metres deep and 9 kilometres wide, formed by the collapse of a seamount half a million years ago (figure 5.1).

As the evacuation of over 8000 people from their homes following devastating eruptions on the Cumbre Vieja ridge demonstrated to the world in September 2021, La Palma is still an active volcanic island. Fortunately, the observatory was not at risk. At a latitude of +29 degrees with ample rainfall and a near-tropical climate, La Palma was the last port of call for many sailors before setting sail for the Americas. There is a lifelike (but, on closer examination, painted concrete) replica of Christopher Columbus's boat, the *Santa Maria*, in Santa Cruz, the largest town. As one of the steepest populated islands on the planet, its narrow roads wind tortuously through evergreen temperate cloud forests interspersed with tiny villages containing whitewashed houses bedecked with vivid red, orange, and purple flowers. Despite this exotic location, La Palma

FIGURE 5.1. The telescopes at the Observatorio de Roque de Los Muchachos on the summit of the island of La Palma are close to the deep crater called the Caldera de Taburiente.

is one of 17 autonomous communities in Spain and thus firmly in the European Union.

The Canary Islands represent the birthplace of the now well-established practice of locating observatories on mountaintops. Newton first contemplated the idea in his treatise *Opticks* (1704), in which, despite never travelling beyond England, he considered the astronomical advantages of the "serene and quiet air . . . above the grosser clouds."[1] However, it was Charles Piazzi Smyth (1819–1900), Astronomer Royal for Scotland, who demonstrated the wisdom of this suggestion via an expedition to Mount Teide, the 3700-metre peak of Tenerife. Accompanied by his wife, assistants, and equipment carried by mules, his colourful account was accompanied by stereoscopic photographs.[2]

1. 4th edition (1730; repr. Dover 1952), book I, p111.

2. *Teneriffe, an Astronomer's Experiment; or, Specialities of a Residence above the Clouds,* Charles Piazzi Smyth (1858; repr. Cambridge University Press 2010).

What is more conveniently referred to as the La Palma observatory began with the controversial relocation of the 2.5-metre Isaac Newton Telescope (INT) discussed in chapter 1. Operated by what was then the Royal Greenwich Observatory (RGO), based at the time in Herstmonceux, Sussex, the refurbished INT was joined—in a partnership with the Netherlands—by the smaller 1.0-metre Jacobus Kapteyn Telescope in 1983 and the larger 4.2-metre William Herschel Telescope (WHT) in 1987. This collection of three telescopes became the first British optical observatory on a high-quality astronomical site in the northern hemisphere. La Palma was a much more convenient destination for British astronomers than Australia, and, unlike the 50:50 arrangement at the AAT, they had the lion's share of the observing time.

Recognising this expansion of UK astronomical facilities, I arranged a leave of absence from Durham University via a research position at the RGO. Our growing Durham instrument group was building a Faint Object Spectrograph (FOS) for the INT and the move would give me time, free from teaching duties, to install this spectrograph and use it to do research. As an RGO employee, however, I would also be a support astronomer at the INT, assisting visiting astronomers during their observing runs. As I briefly discussed in chapter 2, support astronomers are the scientific lifeblood of an observatory, and their duties range from helping visiting astronomers use the complex instrumentation to undertaking observations on behalf of astronomers who choose to wait for predetermined weather conditions. These two modes of observing are usually referred to as "visitor mode" and "service mode," respectively. Most of the RGO astronomers were, like me, obliged to undertake support duties, and a small number chose to live with their families on La Palma. However, owing to the absence of English-speaking schools on the tiny island, this was not a popular option and most, including myself, would fly to La Palma for a tour of duty of 10 nights or so, assisting various groups of astronomers as well as undertaking observations in time the support astronomers had secured for their own research programmes (plate 25).

La Palma is a popular tourist destination, and I would take a 4-hour charter flight from the United Kingdom, which was usually completely full of families and couples headed for a week or two of sunshine on its beaches. On one return trip I recall a plane full of well-tanned travellers, some a delicate shade of lobster. As I had been working at night and sleeping during the day, I was certainly a pale outlier in this holiday crowd. Soon after take-off, a middle-aged woman sitting next to me and noting my white skin whispered gently, "Were you sick, dear?"

The FOS was the first astronomical instrument for which I had a scientific responsibility as a "project scientist." Unlike some of my later instruments, such as the Low Dispersion Survey Spectrograph (LDSS; chapter 4) built for my personal research, FOS was a "common-user" instrument, meaning it was designed to serve the entire astronomical community. I had to oversee the writing of a comprehensive user manual and acted as a point of contact for all queries, technical and scientific. Together with the RGO, we later built a second version (FOS-2) for the WHT; the project scientist on this occasion was Jeremy Allington-Smith, a leading member of the instrument group at Durham.

The FOS, designed by Charles Wynne at RGO in 1982, had two remarkably innovative features (plate 26). A traditional spectrograph gathers light in a converging light beam at the focal plane of the telescope. The diverging beam of light that enters the spectrograph through a rectangular aperture (or "slit") is normally rendered parallel by a "collimator" (either a lens or curved mirror). This parallel beam then passes through a diffraction grating where it is split, or "dispersed," into its colours. Light of different colours continues in a parallel beam, the angle of transmission after dispersal depending on the wavelength. A "camera" mirror or lens brings the dispersed beams into focus on a detector. Since a fraction of light is lost at each air-glass surface, Wynne had the courage to consider what would happen if the collimator optics were removed. In this case, the beam entering the spectrograph would be divergent as it encountered a transmission grating. Although complex optical aberrations would result, Wynne reckoned they could be eliminated with a glass correcting element attached to the grating. The second innovation followed the arrival of compact CCD detectors, which could be placed at focus within the spectrograph without causing any obstruction, avoiding the need for further optics to deflect the beam to an external focus. The resulting spectrograph was the size of a suitcase and astonishingly efficient, ideally suited for faint object spectroscopy.

At the time I had little experience of designing spectrographs. While both FOS-1 and FOS-2 were efficient, only later did I realise they were effective only for a restricted category of galaxies with intense gaseous emission lines. The spectral resolution of the FOS design—that is, its ability to discern weaker features, such as the diagnostic stellar absorption lines of hydrogen, calcium, and sodium—was poor. A traditional spectrograph usually offers a range of diffraction gratings, each of which can be inserted to suit varying scientific applications. However, given its complex design, FOS had a fixed format: its diffraction grating could not be swapped for one with higher resolution.

Fortunately, for dusty star-forming galaxies selected at infrared wavelengths from an all-sky survey conducted by the Infrared Astronomical Satellite, both FOS-1 and FOS-2 were perfectly suited. In a UK team led by Michael Rowan-Robinson, then at Queen Mary College London, we undertook a redshift survey of nearly 1500 dusty galaxies to chart their three-dimensional distribution. It provided us with an early glimpse of large-scale structure in the universe in the late 1980s. Since this survey predated the revolution of multi-object spectroscopy at the AAT (chapter 4), an undertaking in which galaxies are examined one by one could be achieved only with a remarkably efficient spectrograph; on one night I managed to measure over 100 redshifts.

The late 1980s was a time of great progress in UK astronomy. We were by now fortunate to have access to 4-metre-class telescopes in both hemispheres and could look at our European and American competitors with confidence and pride. Although I was still travelling regularly to Australia, I enjoyed observing at La Palma. Returning full time to Durham in 1985 after my 2 years at RGO, I had learned a lot about instrument building and observing. I also knew the La Palma observatory staff very well as I had worked alongside them during my many tours of duty. It was amusing be on this exotic island and enjoy the camaraderie, talking to colleagues who had the full range of regional British accents.

The island also has its considerable charms. Leaving a Durham winter and stepping off the plane onto the airport tarmac in the warm, humid air made me realise how fortunate I was to be an astronomer. The airport at Santa Cruz was quite provincial at the time; even the airport security staff seemed laid back. Returning home on one trip with a box of magnetic tapes containing my precious astronomical data, I warned the guard at the security gate that placing the tapes through the baggage scanner might risk destroying my precious data. This was not recommended, I said, and would he inspect the tapes manually? He declined the offer and urged me to put them through the scanner. Resisting his instructions twice with my primitive Spanish: "Datos magneticos, muy importante, Observatorio de Roque de Los Muchachos!" he eventually winked at me and said quietly, "Machine is not switched on"!

Observing from the summit of La Palma was inspirational. As the afternoon sun sunk over a dark blue sky, I could stare far west over endless miles of the Atlantic ocean, or clearly see the peaks of other islands in the Canarian chain across the abyss of the deep Caldera. While the initial accommodation set up by the RGO in the early 1980s was rather primitive, the IAC eventually completed a residencia with first-rate bedrooms and a canteen providing meals and night lunches that were infinitely tastier than the dull fare at the AAT.

As with the AAT, one taxi driver had the contract to take astronomers to the observatory. A stocky Canarian named Leonel, he drove a white Mercedes that he threw around the endless hairpin bends winding up the mountain. Many astronomers dreaded this ride and got carsick in the process. During the winter months, portions of the road occasionally got washed away or were covered by a thin layer of ice at dawn, but fortunately we were in capable hands with Leonel. Although his English was minimal, he somehow absorbed all the "astro-gossip," presumably from listening to the astronomers' banter in his taxi. At a time when I was contemplating a move to the United States (chapter 7), he once picked me up at the airport and, no sooner than I'd sat down in the front seat, stared me in the face and asked in his broken English, "You moving to California or you going to Hawaii?"

Returning to the study of high-redshift galaxies, I think it's fair to say that my own research progress was still determined mostly by the work I was doing at the AAT. Since the WHT didn't become available until

1987, inevitably all the instrument development we had initiated at Durham relating to multi-object spectroscopy (Parry's automated fibre positioners; the LDSS with Taylor and Atherton) was completed and exploited initially on the AAT. Although, in competition with efforts at Palomar, I undertook studies of distant clusters of galaxies with FOS-2 on the WHT out to redshifts of $z \sim 0.6$, it was hard going given the low spectral resolution of that instrument.

The original intention with LDSS (chapter 4) was that it would be used initially on the AAT but transferred to the WHT when the latter telescope became available. However, the AAT director, Don Morton, objected to losing our instrument once he saw how productive it was. He rightly argued that the observatory had invested much effort in helping us to install it. As a cheap, manually controlled instrument, it was also unclear how we could use it on the WHT, which had no Cassegrain cage in which night-time observers could ride. Accordingly, in the late 1980s Keith Taylor and I convinced the RGO to fund a second, fully automated version, dubbed LDSS-2, which would be a common-user instrument available for the community. At Durham, Jeremy Allington-Smith, who had done a great job on FOS-2, became the LDSS-2 project scientist and took the project to its completion by 1992. I have a soft spot for LDSS-2 because a refurbished version of the instrument is still in active use at the Las Campanas observatory in Chile (plate 27).

In 1993 I left Durham, after 19 years, and moved to Cambridge. Although apprehensive about leaving given how well things were going, I could not turn down the historic Plumian Chair of Astronomy and Experimental Philosophy, which dates back to 1704. Previous holders included such luminaries (and my heroes) as Arthur Eddington, Fred Hoyle, and Martin Rees. I was particularly sad to leave the Durham instrumentation group, which had produced four spectrographs (FOS-1, FOS-2, LDSS-1, and LDSS-2) and initiated the innovative fibre-positioning system Autofib. The development of each instrument had been driven by my research group's desire to secure redshifts efficiently for successively fainter galaxies. Fortunately, however, four of my research students and postdoctoral researchers decided to move with me to Cambridge, which helped me to settle in quickly. A very talented postdoc who got his PhD in Edinburgh, Karl Glazebrook, led the first substantial effort with the newly installed LDSS-2 on the WHT.

The explanation for the excess number of faint blue galaxies (introduced in chapter 4), remained a topic of intense discussion at the galaxy conferences I attended in the early 1990s. From our AAT surveys, Tom Broadhurst, Matthew Colless, and I had argued that the excess was likely caused by the more vigorous star formation seen preferentially in low-luminosity galaxies at earlier times. However, some astronomers proposed other explanations. Len Cowie, a Scotsman based at the University of Hawaii (plate 28, *right*), and colleagues suggested that the number of galaxies per unit volume in space might not be conserved over time. A present-day galaxy might be the product of the merger of earlier, smaller systems. When looking back in time one would then see an excess number of galaxies without necessarily probing further in distance than expected for a non-evolving population. However, the fact that the count excess seemed to be associated with strong gaseous emission lines—indicators of active star formation—favoured our hypothesis. A more radical idea proposed by a Japanese astronomer, Yuzuru Yoshii, suggested that we had underestimated the volume in which galaxies were being counted. He proposed adjusting the cosmological model; an idea that seemed unnecessarily revolutionary to me. Meanwhile, David Koo and colleagues continued to promote Richard Kron's idea that the excess could arise from an underestimated number of nearby feeble "dwarf" galaxies.

Glazebrook and I reckoned that, with the better observing conditions on La Palma and the improved performance of LDSS-2, we could push quite a bit deeper than had been done in the survey Matthew Colless and I completed with LDSS-1 on the AAT. Cowie had already made a modest step in this direction by securing the redshifts of 12 galaxies beyond the LDSS-1 limit using the Canada-France-Hawaii Telescope (CFHT) on Maunakea. He also demonstrated the important role of infrared images of faint galaxies, arguing these were a valuable indicator of their collective mass in long-lived stars suitable for testing his merger hypothesis. With LDSS-2 we made a solid advance over Cowie's work, securing the redshifts for 73 galaxies beyond the LDSS-1 limit, probing four times fainter than our earlier work. In terms of record breaking, we reached beyond a redshift $z \sim 1$ for the first time, corresponding to a lookback time of over 8 billion years (60% of the way back to the beginning of the universe).

There were several novel aspects of this study. Foremost, unlike the AAT LDSS-1 campaign, our selection was based on CCD images rather than photographic plates, reflecting the technical progress in detectors over the previous decade. Secondly, since at these faint limits one cannot easily distinguish visually between stars in the Milky Way and external galaxies, we chose to secure spectra for all objects down to our brightness limit. This was not a great overhead since, compared with 73 galaxies, we found only eight stars. This minimal contribution of foreground stars emphasised that we were finally probing sufficiently faint that most sources were external galaxies. (This was certainly not the case in our earlier AAT surveys, where the number of stars was comparable to that of galaxies and great care was needed to remove stars from the target list by analysing their compact morphologies on the photographic plates.)

This survey reached the faintest limits I ever accomplished with the United Kingdom's telescopes. Glazebrook and I undertook marathon exposures of up to 6 hours on the WHT in four deep fields. Although we broke the "redshift one barrier," the mean redshift of our LDSS-2 sample was still only $z = 0.46$, consistent with the no-evolution hypothesis even though the excess numbers at this new limit were by a factor of almost four (figure 5.2). The continued absence of a tail of low-redshift ($z < 0.2$) sources firmly ruled out a population of undiscovered local dwarf galaxies. The absence of a prominent tail of high-redshift sources ($z > 1$) ruled out evolutionary brightening shared by the entire galaxy population.

Reflecting now on this period in my career I believe that, pioneering though our surveys were in pushing back the frontiers, we were simply reiterating (albeit with increased confidence) the story of luminosity-dependent evolution promoted 7 years earlier in the article with Tom Broadhurst. On the one hand, after this huge effort, the mean redshift of the LDSS-2 survey published in 1995 ($z = 0.46$) did not represent much of an advance over the original fibre-based Broadhurst survey of 1988 ($z = 0.25$). The scientific conclusions were the same: the population of

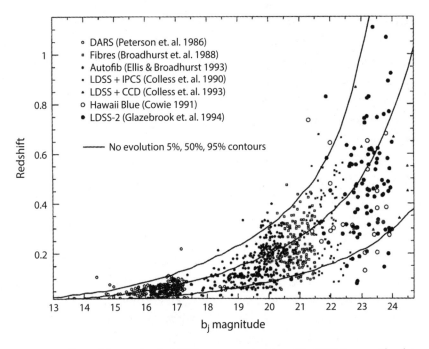

FIGURE 5.2. State of the art after the LDSS-2 survey undertaken at the WHT in 1994. The plot shows the redshift (*y* axis) against blue magnitude, a logarithmic measure of brightness decreasing to the right (*x* axis), for over 700 galaxies drawn mostly from the AAT and WHT studies, the most distant now lying beyond a redshift 1. Despite the excess numbers compared to non-evolving models, the redshift distribution tracks the no-evolution curve (central solid line) remarkably well.

galaxies had evolved over the past 8 billion years, but the greatest change was a declining rate with time of newly formed stars in sub-luminous galaxies. On the other hand, science often progresses in this incremental way. In 1988, this conclusion had not been universally accepted; by probing deeper where the excess numbers were much more significant, we had now demonstrated our original result beyond much doubt. Then, in 1995 a rival team emerged that not only confirmed our original conclusion but took the subject further through a more ambitious and thoughtful strategy.

Simon Lilly is the epitome of a foreigner's view of an Englishman. A Cambridge graduate, well-spoken and smartly dressed, he sported a tidy moustache reminiscent of a wartime pilot in the Royal Air Force (plate 28, *left*). After a spell at Princeton and a brief period collaborating with Len Cowie in Hawaii, he moved to the University of Toronto and began the Canada-France Redshift Survey (CFRS) using multi-slit instrumentation on the 3.6-metre CFHT on Maunakea. He teamed up with an equally ambitious Frenchman, Olivier Le Fèvre (1960–2020), based in Marseille, and together with David Crampton, a Canadian at the Herzberg Institute of Astrophysics in Victoria, British Columbia, and the Frenchwoman Laurence Tresse (who later became director of the Astrophysics Research Centre at Lyon), embarked on a large red-shift survey, pooling observing time from both the Canadian and French time allocation committees.

Lilly thinks deeply about his science and has strong leadership and management skills, and he and his colleagues rapidly executed a formi-dable observing survey, bringing several new ideas to the fore. Recognis-ing the importance of having infrared data, the CFRS survey began with a multicolour imaging campaign in five deep fields. Whereas my team had consistently selected galaxies imaged through a blue filter sensitive to recent star formation, the CFRS team argued this strategy might lead to unnecessary uncertainties of interpretation. For blue-selected targets, the light that left high-redshift galaxies would be emitted at a shorter wavelength in the ultraviolet, a wavelength range where, at the time, the properties of local galaxies were poorly understood and comparisons would be tricky. Moreover, our blue selection procedure would bias us toward finding star-forming galaxies at the expense of understanding evolution in quiescent (non-star-forming) examples. Although we ac-counted for this bias in our model predictions, Lilly and colleagues ar-gued that our deductions would be sensitive to the assumptions made. Accordingly, the CFRS team decided to select its galaxies using a far-red filter, thereby securing a more representative mix of both the star-forming and quiescent population out to large redshifts.

A more fundamental advance was the sheer scope of the CFRS survey. The AAT and WHT surveys we had undertaken over 7 years produced redshifts for about 370 faint galaxies. Within just 2 years of intense ob-servations at CFHT, Lilly and colleagues secured spectra for nearly 600 galaxies to an equivalent brightness limit (figure 5.3). Whereas we

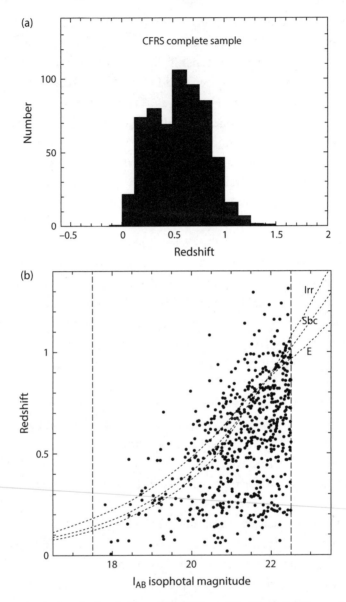

FIGURE 5.3. Results from the Canada-France Redshift Survey. (*Top*) Redshift histogram from five survey fields; (*bottom*) redshift versus brightness for all galaxies (brighter objects to the left of the plot). Through a concerted effort, in only 2 years Simon Lilly and colleagues secured 50% more redshifts than our collective achievements at the AAT and WHT over 7 years and, via a red selection of targets, pushed to substantially beyond a redshift $z \sim 1$.

penetrated the "redshift one barrier" with just a couple of galaxies, the CFRS team, by virtue of its red selection, was able to locate 25 $z > 1$ galaxies. It published its results in an impressive series of 14 highly detailed articles from 1995 to 1997, which had a large impact in the astronomical community. Scarcely a month went by when there wasn't yet another paper by the CFRS team! By virtue of the group's different selection method, it demonstrated that there was little evolution in the quiescent population, but confirmed our finding that evolution in the star-forming population was largely occurring in the low-luminosity systems. In a summary paper the team claimed (rather proudly) that "the picture of galaxy evolution presented here presents no inconsistencies with the very much smaller samples . . . selected in other wavebands" (i.e., our modest efforts!).[3]

The CFRS survey demonstrated to many of us what could be accomplished by a large, well-coordinated team. Observational astronomy was, in the 1990s, transitioning from projects undertaken by traditionally small teams such as mine—typically a professor, students, and postdocs in one institution—to those involving a consortium of leading astronomers spread across a number of institutions. While the CFRS team had only five senior players, ambitious surveys would soon involve teams of 20 senior astronomers or more. Although I fought this trend as long as I could, preferring to work only with colleagues in my own institution (or former members of my group who had recently moved elsewhere), the trend to bigger teams has continued. Many observatories, including Hubble, Spitzer, and ground-based facilities run by the European Southern Observatory (chapter 9), now regularly solicit what are termed "legacy," "treasury," or "large" proposals, promising over 10 times the traditional allocations of observing time, sometimes spread over 2 or 3 years, in order to encourage ambitious science programmes. This has led to the formation of large multinational teams where each participating member has a predetermined role in the data processing

3. The Canada-France Redshift Survey. VI. Evolution of the Galaxy Luminosity Function to $z \sim 1$, S. J. Lilly, L. Tresse, F. Hammer, D. Crampton, and O. Le Fèvre, *Astrophysical Journal*, vol 455 (1995), p123.

and/or the scientific exploitation. It is not always easy to ensure that students and postdocs—those similar to Tom Broadhurst, Matthew Colless, and Karl Glazebrook—can demonstrate their skills and thereby establish their careers in such large teams, which are nearly always led by senior astronomers safeguarded by permanent positions. Admittedly, however, a new generation of astronomers has emerged that is enthusiastic about participating in such large teams, which, if properly managed, can both be more inclusive and offer unique opportunities to share in important discoveries.

Unable to compete with the CFRS survey at the faint end, I thought it would be advantageous to increase the galaxy redshift sample at intermediate brightness, extending our earlier surveys over wider areas of sky. This would reduce the effect of clustering patterns in the galaxy distribution and improve the range of galaxy luminosities sampled at each redshift. Since the Autofib fibre positioner on the AAT encompassed a field of view on the sky over 10 times larger than the multi-slit spectrographs at the CFHT and WHT, it was the ideal instrument for this task. Over a number of runs in Australia, we secured over 1000 redshifts in 32 independent fields at depths similar to Tom Broadhurst's original fibre survey. Together with the fainter surveys, this led to a new catalogue of over 1700 redshifts spanning an unprecedented range in brightness of a factor 100,000. Our resulting article was published at more or less the same time as the final CFRS papers and similarly concluded that there was no change in the quiescent population to at least a redshift $z \sim 0.5$, but that there was a significant increase in activity in the sub-luminous star-forming population. We concluded that "the steepening of the [faint portion of the] luminosity distribution with lookback time is of the form originally postulated by Broadhurst, Ellis and Shanks and is a direct consequence of the increasing space density of blue star-forming galaxies at moderate redshifts" (figure 5.4).[4]

4. Autofib Redshift Survey. I. Evolution of the Galaxy Luminosity Function, R. S. Ellis, M. Colless, T. Broadhurst, J. Heyl, and K. Glazebrook, *Monthly Notice of the Royal Astronomical Society*, vol 280 (1996), p235.

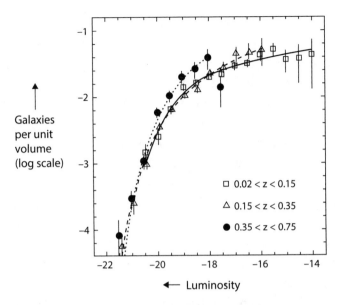

FIGURE 5.4. The wide range in brightness probed over the redshift range $0 < z < 1$ via the extended Autofib redshift survey enabled computation of the luminosity distribution at various epochs. The plot shows the volume density of galaxies (y axis) versus luminosity (x axis, increasing to the left). The faint end slope of this luminosity function steepens at higher redshifts. Less luminous sources are more abundant at higher redshift, thereby producing the excess numbers seen in the galaxy counts.

Allan Sandage was always a strong supporter of my spectroscopic studies of faint galaxies. During one of my frequent visits to Pasadena in the 1990s, I recall him confessing that he had always hoped the Hale 200-inch telescope would have been at the forefront of faint multi-object spectroscopy. Although Caltech did eventually build a fibre-based spectrograph, its performance never seemed particularly impressive, and it made only a modest impact compared with the CFHT, AAT, and WHT programmes. Sandage served on the editorial board of the prestigious journal *Annual Reviews of Astronomy and Astrophysics*, and in 1995 he invited me to write a review of the recent progress made in galaxy studies, both with the various spectroscopic surveys and the first Hubble images of high-redshift galaxies (chapter 6).

Writing such a review article is a major task. Its aim is to introduce the relevant scientific endeavour and goals, critically summarise the methods in current use, detail the recent progress and uncertainties, and suggest directions for the future. Inevitably, many astronomers are eager to see whether their contributions are favourably mentioned. Others are furious if their work is criticised, misrepresented, or, worse, totally ignored. As a result, an author takes on quite a responsibility. *Annual Reviews* publishes typically 15 to 20 articles across all areas of astronomy every year. Most are multi-author affairs, and there was initially a suggestion I team up with Chuck Steidel at Caltech (c.f. chapter 6), but that particular partnership didn't work out and I ended up proceeding on my own.

At the time I was director of the Institute of Astronomy at Cambridge and had responsibility for a department with over 150 astronomers and administrative staff. Working on this review in my office was a hopeless proposition given the endless interruptions. First, I tried to hide away, by writing at home in the afternoons. But later I accepted an invitation to visit the University of California at Berkeley. I was accommodated in the Women's Faculty Club, of all places, but spent most of my visit unsociably in my room, poring over dozens of scientific papers. Neither laptops nor the internet was very effective at that time, and so I was carrying around a heavy box of printed articles marked up with comments and writing my review longhand in a notebook. I failed to meet the initial deadline set by the journal but was fortunately able to negotiate an extra year. So it was with a huge sigh of relief that I submitted a 55-page article entitled "Faint Blue Galaxies" in late 1996.[5]

My joy in completing the review was short-lived, however. Soon after, I gave a copy to a visiting Japanese astronomer, well known for his curmudgeonly behaviour, and, true to form, he came to my office the next morning holding the manuscript, stared me in the face, and said "Frankly, disappointing"! Fortunately Allan Sandage disagreed. He was the principal reviewer of my article and his annotated version was rich in positive comments and suggestions. I later learned that, as one of only three editors of the journal, he would similarly annotate half a dozen or more such reviews each year.

5. Faint Blue Galaxies, R. S. Ellis, *Annual Reviews of Astronomy & Astrophysics*, vol 35 (1997), pp389–443.

My "Faint Blue Galaxies" review was well timed, particularly as the first deep images from the Hubble Space Telescope graphically illustrated how dominant these blue sources are at faint limits (plate 29; more on this in chapter 6). Much of my article discussed the spectroscopic surveys undertaken by our group and the CFRS team. Areas that I considered promising for the future included: (1) deeper imaging with the Hubble Space Telescope, which offered the opportunity to reveal the morphological forms of blue star-forming galaxies; (2) using gravitational lensing to magnify the light of distant objects seen behind foreground massive galaxies or clusters (discussed further in chapter 7); and (3) the use of "photometric redshifts"—multicolour data to estimate approximate redshifts for very faint sources well beyond the spectroscopic limit (chapter 8).

Reading my article 25 years later it comes across as remarkably primitive, which, I suppose, simply emphasises the dramatic progress we've achieved subsequently, which is highlighted in the remaining chapters. There is no mention of "cosmic dawn" or reionisation (chapter 8), and the article is preoccupied with discussing how galaxies evolved since a redshift $z \sim 1$–2, at a time where it was claimed the bulk of star formation occurred. It seems that a redshift $z \sim 2$–4 was honestly regarded as the horizon in the 1990s. Nonetheless, the areas of future promise I discussed have indeed flourished, so I think the review was reasonably insightful. Years later I met a professor of English literature who was visiting Caltech, and when I mentioned how hard I had worked on my review only to see how quickly it became out of date a couple of years later, he told me how lucky I was to work in such a field. He claimed that research in his field had been stagnant for over a decade.

Before we leave La Palma, I would like to change topics and describe a very important decision that reshaped the United Kingdom's relationship with the island in relation to plans for future large telescopes. As mentioned earlier, I served on a Royal Astronomical Society committee in 1985, which produced a report recommending future observing facilities for UK astronomy. One of its priorities was a wide-field spectroscopic capability, which ultimately led to my proposal to the AAT board of directors for the two-degree field facility (chapter 4). However, the highest priority item in the report was for Britain to gain access to an 8-metre-class telescope. By the mid-1980s, Caltech and the University of California were planning what became the Keck Observatory (chapter 7). Likewise, the ESO, the US National Optical Astronomy Observatory (NOAO), and the Japanese each had plans for similar large facilities. Although the United Kingdom was now in a strong position internationally, with generous access to a suite of three 4-metre telescopes—the 3.9-metre AAT, 4.2-metre WHT, and the 3.8-metre UK Infrared Telescope (UKIRT) on Maunakea—the RAS report argued the country would fall behind on the international level if it did not also develop plans to gain access to an 8-metre-class telescope.

Technologically, the time was ripe for considering a new generation of telescopes. Larger mirrors were now considered a practical proposition, either via a mosaic of carefully controlled hexagonal segments (the Keck approach) or through thin meniscus-like mirrors actively supported to maintain their shapes (as demonstrated at UKIRT and later by the ESO New Technology Telescope). Primary mirrors with shorter focal lengths meant smaller and cheaper telescope enclosures, and large (so-called Nasmyth) platforms, such as those employed at the WHT, could accommodate large scientific instruments. The United Kingdom had experienced engineers and opticians in each of these areas. The main challenge in moving forward was, of course, raising the funds. As Britain was not then a member of ESO, it seemed likely it would have to find an international partner to share the cost.

In 1987 the UK Science and Engineering Research Council (SERC) asked me to chair its Large Telescope Panel (LTP), whose charge was to assemble the scientific case for an 8-metre-class telescope and examine both the technical issues and international partner options. The panel included two other university-based astronomers, Mike Edmunds at Cardiff and Jim Hough at Hatfield, as well as astronomers and technical experts from the two royal observatories and the Rutherford Appleton Laboratory. The pace of meetings was very fast. We were charged with reporting within 6 months.

One might imagine today that the scientific case for a more powerful, 8-metre-class telescope was self-evident but, as always, there were mixed

views. Our case was assembled with input received from many individuals in the community who were experienced in various areas of
observational astronomy. Our panel then toured the country in a series
of "roadshows," presenting the case and rallying support. During these
many presentations, I was often heckled and fiercely criticised. Many
astronomers thought it was too soon to embark on planning for a new
facility; after all, the 4.2-metre WHT had been completed only that year.
Others thought the plan was insufficiently adventurous and that the
United Kingdom should "skip the 8m club" and wait for a future era of
16-metre-class telescopes. As always, some were concerned the cost of
a new facility would divert funds from their area of astronomy and even
lead to closure of their favourite existing telescope.

The most surprising reaction came from a well-respected X-ray astronomer, Andy Fabian at Cambridge, who argued that for the price of
a single 8-metre telescope, the United Kingdom could instead buy four
more 4-metre telescopes, thereby significantly increasing the number
of observing nights for the community. I simply disagreed with this
suggestion, since I believed it wouldn't lead to any advance in observing
capability. Moreover, at a practical level, building four new telescopes
and operating them (in addition to the three already available to UK
astronomers) would present an extraordinary challenge in terms of locating the necessary workforce. To my amazement, however, Fabian's view
received a lot of support. It seemed many astronomers feared observing
time on a single 8-metre telescope would be highly competitive and
they would miss out, beaten as it were by a select elite group. Better to
provide more nights on many 4-metre telescopes so everyone could get
a share. Behind the scenes I feared some were saying that Richard Ellis,
a faint object astronomer, was simply promoting his own facility.

Although our report was welcomed by the majority of the community, SERC was under immense financial strain from the late 1980s to
the mid-1990s. This arose partly from government policy in the Margaret Thatcher era and was one of the reasons why the British brought in
the Dutch as 20% partners to complete the WHT. Within some sections
of SERC, our push for an 8-metre telescope was viewed with incredulity.
UK astronomers now had access to the AAT, WHT, UKIRT and the

James Clerk Maxwell millimetre telescope on Maunakea. "How much more do they want?" grumbled some in authority. Nonetheless, there was reassuring support from Ian Corbett, SERC's head of Astronomy and Particle Physics, and, in 1989, the case for 8-metre-telescope access was strongly supported in a future-looking document, the SERC "Ground-Based Plan." This led to funding for a more detailed 2-year study of the likely capital and operational cost of a 50% share of an 8-metre telescope, and a British astronomer, Roger Davies, who had been working in Arizona, was recruited as project scientist and established a team in Oxford.

The most interesting aspect of this story, and where La Palma features prominently, relates to finding an international partner. The LTP began this adventure immediately, prior to the arrival of Roger Davies. Today, I find it amazing that I was permitted, aged in my late 30s, to travel the globe (with other panel members) to investigate possible international partnerships on behalf of the United Kingdom. We often used to begin our overseas presentations with a slide entitled "UK Large Telescope Marriage Brokers." Four options were explored, and these were narrowed down to two by the time Roger Davies arrived in post.

My first visit was to the ESO headquarters near Munich in 1987. With Alec Boksenberg (director of the RGO) and Alf Game (an SERC employee and secretary of the LTP), we met the director-general of ESO, Lodewijk Woltjer (1930–2019). ESO was planning its Very Large Telescope (VLT), an array of four 8-metre telescopes in Chile. One option we discussed was UK funding, perhaps with Spain, of a fifth 8-metre telescope of ESO's design to be placed on La Palma. The ESO community, together with the United Kingdom and Spain, would then share access to four telescopes in the southern hemisphere and one in the north. Woltjer did not like this proposal, arguing it would burden ESO in additional negotiations and geographical complications. At the time the VLT was not yet fully funded, and he preferred that Britain simply join ESO and participate in the VLT project. As the United Kingdom was not a founding member of ESO, it would have to make additional contributions to join the organisation. Woltjer suggested that the AAT would do nicely! Given that the AAT had, in many respects, catapulted UK astronomers ahead of their ESO counterparts, Boksenberg and I regarded Woltjer's counterproposal almost as an insult. We had great pride in the independent strength of the United Kingdom in observational astronomy and left Munich without a desire for proceeding further.

Interestingly, the idea of a fifth telescope on La Palma was resurrected at the initiative of Woltjer's successor as ESO director, Harry van der Laan, in 1989. The VLT was now an approved project, and ESO was about to place its contracts for the four VLT telescopes in 1989–1990. If the United Kingdom ordered a fifth telescope, there would naturally be some cost saving. However, since the UK telescope would be the last to be constructed, we would be at the mercy of ESO's own timetable. Together with the fact that UK industry wouldn't necessarily get any construction contracts, this option no longer appealed to my panel. This particular story ends when Britain did, eventually, join ESO 15 years later in 2002. Its additional contribution at the time was a new 4.1-metre telescope called VISTA (the Visible and Infrared Survey Telescope for Astronomy).

Our next trip was to Japan. Following an initial scouting trip by Jim Hough, the LTP travelled to Tokyo, where we met our Japanese counterparts at what was then Tokyo Observatory in the suburb of Mitaka. In a large room at the observatory, about 20 Japanese astronomers sat on one side of an enormous polished table, and the Brits were invited to sit along the other side. Staring across the table, the age difference was striking. The director of Tokyo Observatory, Yoshihide Kozai (1928–2018), was about 60 and surrounded by similarly aged colleagues, whereas most of the Brits were in their 30s or early 40s. This was my first experience of dealing with senior Japanese astronomers. Although we were warmly welcomed, they seemed cautious and mindful of staying on good terms with their science ministry. They hoped to place a 8-metre telescope, the Japanese National Large Telescope in Hawaii, but worried that their ministry had no experience of significant investment outside Japan. They feared that a funding partnership with the United Kingdom would present an even more difficult task for their higher-ups to manage.

That evening we were entertained at an exclusive Tokyo restaurant. The meal consisted of 10 courses, delivered together in a pyramid-shaped tower of porcelain bowls. Each layer in the pyramid contained a different delicacy. We proceeded to eat for over an hour or more, layer by layer, from top to bottom. The final course presented a mystery. The lowest porcelain bowl was, naturally, the largest and had a small circular hole in its lid. Peering inside with some anticipation, the contents appeared to be simply clear water. However, when the lid was removed, floating in this water was a tiny half-mushroom. Some of my British colleagues expressed surprise, even amusement, at the apparent anticlimax. At this point, my Japanese neighbour leaned over to me and said "Special mushroom, grows in pine forest only certain time of year, 100 dollars!"

The visit was, nonetheless, very fruitful. It was clear the Japanese did want some form of deal, perhaps reflecting their admiration for the achievements of UK astronomy, which they sought to replicate. Largely at their request, we signed an agreement to cooperate both technically, on telescope designs and astronomical instrumentation, and scientifically, through joint programmes on La Palma funded in part by the Royal Society. For the next two decades, the UK and Japanese communities worked together very well. The Japanese National Large Telescope was eventually completed in 1998 and renamed the Subaru Telescope. Shortly thereafter, it was opened officially by Princess Sayako, the youngest child of Emperor Akihito. One of the science instruments on Subaru involved a major contribution from Oxford University.

With the growing assemblage of UK-funded telescopes on La Palma and the central role of the RGO in their operation, there was naturally strong pressure to realise an 8-metre telescope on the island, logically in a partnership with Spain, which, at the time, was also not a member of ESO. This "Spanish option" became one of the two prime routes forward for the LTP to evaluate. The Observatorio de Roque de Los Muchachos was managed by the IAC, an institute established in 1975 by founding director Francisco "Paco" Sánchez, a warm and friendly man who can truly claim to have brought Spanish astronomy into the modern era.[6] As is common practice for observatories where a local organisation

---

6. An English translation of Sánchez's colourful, and in places heart-warming, account of his leadership of Spanish astronomy is available in *The Rise of Astrophysics in Modern Spain: From Dictatorship to Democracy* (Springer 2021)

provides and maintains the site infrastructure, such as summit roads and security, there is a local tax on observing time. Although the INT and WHT were mostly UK funded and operated by the RGO, the IAC received 20% of the observing time on each. Initially, the SERC was miserly in response to my panel's report, complaining that it was unlikely half an 8-metre telescope could be afforded by the United Kingdom. Given ESO was proposing a facility comprising four 8-metre telescopes, it was depressing indeed that Britain could not muster the resources even for half of one. As a result, further discussions were held with several other European countries to make up the difference. In the end, Sánchez proposed not only to match the UK contribution to an 8-metre telescope but also to waive the 20% tax. Much to my relief, the SERC then rose to the challenge of considering half the necessary funds.

The "American option" can be traced back to the plans of the federally funded NOAO, whose headquarters are in Tucson, Arizona. In the mid-1980s, NOAO was promoting a National New Technology Telescope (NNTT), an ambitious facility consisting of four 8-metre mirrors on one mount, equivalent in light-gathering power to a single 15-metre telescope. Roger Davies, who became the UK Large Telescope project scientist, had played a role in assembling its scientific case. Given NOAO's charge is to serve the entire US astronomical community, one might imagine such a bold step forwards would be appropriate. After all, in some respects the NNTT would be a comparable facility to ESO's VLT.

However, those US universities with privately funded telescopes, such as Caltech, the University of California, and the University of Texas, were often wary of NOAO overreaching itself, lest significant national resources be diverted away from other projects, including instrumentation for their own facilities. As a result, senior figures in private US universities were often quick to criticise NOAO's plans. In the end, the NNTT project was shelved as being overly ambitious and costly.

It was at this time I met the outgoing director of NOAO, John Jefferies. I was attending a meeting on multi-object spectroscopy at Tucson, and, in the margins of that meeting, we had lunch. An Australian solar astronomer, Jefferies had formerly been the director of the Institute for

Astronomy in Honolulu for 16 years, during which time he oversaw the impressive development of optical astronomy on Maunakea. I mentioned my panel's search for a partner for an 8-metre telescope, and he welcomed the initiative, regretting the demise of the NNTT and responding warmly to the idea of a US-UK partnership. In 1987, his replacement as director of NOAO was Sidney Wolff, also a solar astronomer previously in Hawaii. Wolff was the first female astronomer in American history to head a major observatory and became a wonderful force in the story that follows. Matthew Colless, who later became my postdoc working on LDSS-1 (chapter 4) was working at NOAO at the time. He told me that on her very first day as director, Wolff asked him for my phone number in Durham. She called me and expressed great enthusiasm for a partnership with the United Kingdom and immediately wrote accordingly to SERC. Soon after, a delegation of Americans, including Wolff and the president of the Association of Universities for Research in Astronomy (AURA), Goetz Oertel, came to London to meet the LTP and SERC and encouraged us to join with the United States in an 8-metre-telescope partnership.

AURA operates both ground- and space-based telescopes for the US community. In doing so it charges a management fee for both the National Science Foundation (NSF) and NASA, respectively. As a 14-year-old, Goetz Oertel (1934–2021) escaped the advancing Russian army in what is now Poland and was educated in West Germany. Originally a solar physicist, he moved through various senior management roles in astronomy and nuclear energy before becoming president of AURA in 1986. I was greatly impressed with Oertel's statesmanlike behaviour at our meeting in London. Unlike some senior administrators in SERC, he listened patiently to the UK astronomers. After our cordial and constructive discussions, he invited the Large Telescope Panel to drinks at a London club that had reciprocal arrangements with his Cosmos Club in Washington, DC. The AURA-LTP meeting went far better than I ever imagined.

It seemed surprising to some that the United States felt the need for an astronomical partnership with the United Kingdom. But I soon learned that this was part of a new direction following the demise of the

NNTT. Oertel once said that national centres such as NOAO were greatly constrained in their future plans. Viewed from universities such as Caltech and the University of California, a national centre "burning up money" was regarded as a competitor.[7] As incoming president, Oertel saw the benefits of AURA having a more international outlook.

There was also a new push for international cooperation at the highest level of NSF. Following a meeting of the heads of many national research councils in Venice in 1988, Eric Bloch, the NSF director, set up an international joint working group in astronomy with members from Canada, West Germany (at the time not reunited with the East), France, Italy, Japan, the United Kingdom, and the United States. Malcolm Longair (director of the Royal Observatory, Edinburgh) and I were appointed as UK members. The US chair was Laura "Pat" Bautz, director of NSF's Division of Astronomical Science. The charge to this working group was certainly ambitious. We were told to compare our national scientific plans for future astronomical facilities and consider how they might be expanded or realised more rapidly via international collaboration. As our meetings developed, it was remarkable to see the variety of responses to this bold invitation from the various national members. As the LTP had learned, the Japanese were reluctant to play ball as their science ministry was not yet ready for international partnerships, while the continental Europeans were sceptical of this American overture and wrote among themselves to encourage European solidarity; in this pre-Brexit era, Longair and I were included in this correspondence.

However, Longair and I were sincere admirers of US astronomy. Malcolm Longair, originally a Cambridge radio astronomer, was a powerful figure in UK astronomy. He was the Astronomer Royal for his native Scotland and widely respected internationally, from the former Soviet Union, where he had worked alongside famous cosmologists, to the United States, where he was strongly involved in the Hubble Space Telescope. I, likewise, was a big fan of the US astronomical community

7. Oral history interview of Goetz Oertel by Patrick McCray (2001); interview transcript archived at the American Institute of Physics and available at https://www.aip.org/history-programs/niels-bohr-library/oral-histories/24721-1.

following years of observing with UKIRT on Maunakea and visits to many of the major US universities. I liked the vibrancy and the "can do" attitude and thought the United Kingdom would benefit by having its eyes opened. Moreover, I was convinced, given our strength at the time, that the UK community was a match for that of the United States, both intellectually and technically. The end product of both Sidney Wolff and Goetz Oertel's visit to London and the report of the NSF's Joint Working Group was the "American option": two identical 8-metre telescopes, one in each hemisphere, involving the United States, the United Kingdom, and Canada. Britain and Canada would each have 25% of the observing time on each telescope, with the United States taking the remaining 50%.[8]

Together with the LTP, Roger Davies's project office at Oxford focused on evaluating this "American option" alongside the "Spanish option" at La Palma, where, to entice us, Sánchez now promised the United Kingdom 55% of a joint 8-metre. The choice of which option to recommend involved scientific, funding, and political aspects. The American option would place the northern telescope on Maunakea, a widely respected site on US soil. Through our involvement in UKIRT, we knew this to be a high-altitude, cold, and dry site, particularly well-suited for infrared studies. The LTP determined that Maunakea was superior in quality to La Palma, which suffered from warmer temperatures in summer that would increase the infrared background. Our American colleagues were also racing ahead with this twin telescope option, later named the Gemini Observatory, at the suggestion of the Canadian astronomer Gordon Walker. There was a clear sense of purpose and momentum in this option, whereas scientific discussions with the Spanish remained more nebulous. Many senior astronomers in the United Kingdom felt the lack of scientific input from our Spanish colleagues was actually an advantage: Britain would be the dominant partner and get its way—for example, on the choice of instrumentation. Better to be in

8. A full account of the development of what became the Gemini Observatory is in *Giant Telescopes: Astronomical Ambitions and the Promise of Technology*, W. Patrick McCray (Harvard University Press 2004).

control rather than have to deal with unsympathetic Americans, they argued, who had little time for international give and take. This concern about being unable to hold our own with the Americans surprised members of the LTP, given our cordial discussions thus far. Nonetheless, as there had already been significant investment in La Palma, there was widespread support for the Spanish option from many of the most important figures in UK astronomy, such as Alec Boksenberg at the RGO and Martin Rees in Cambridge. Meanwhile, we on the LTP argued for the American option. In addition to the superiority of Maunakea for near-infrared studies, the additional access to the southern hemisphere would enable continued synergies with research on the AAT.

There was widespread debate in the UK community about the two options, which, at times, got quite heated and bitter. A cartoon in the influential magazine the *Economist* summarised the choice under an article entitled "Spoilt for Choice" (plate 30, *left*). Although the LTP's recommendation was eventually endorsed by the SERC ground-based planning committee, it faced a formidable final hurdle at SERC's council. It was widely suspected that the SERC chairman, Sir William "Bill" Mitchell, was opposed to the American option. However, following a tense debate, in 1990 the LTP recommendation for the American option was endorsed, thanks to strong support from Malcolm Longair; Arnold Wolfendale, my senior colleague at Durham (chapter 4); and Ian Corbett, who worked extensively behind the scenes. Wolfendale stressed the importance of the common scientific goals evident in the American partnership, as well as the fact that any future cost escalation would not fall entirely on the United Kingdom. To his credit, Wolfendale changed his intended vote away from the Spanish option only a week before the council meeting, following inconclusive discussions relating to financial commitments SERC held with Paco Sánchez and Spanish colleagues in Madrid. With funding finally approved in 1994, the United Kingdom joined the Gemini project, and the twin telescopes were completed in 2000 (plates 30, *right*, and 31). During this decade of planning and construction, I served with Ian Corbett on various international Gemini committees, including its Science Committee and board of directors. Yet, remarkably, despite my efforts to realise UK membership over

15 years, I have never personally used either Gemini telescope. I was greatly disappointed when the United Kingdom eventually withdrew from the project because of a funding crisis in 2007.

I often wonder what would have happened if so many senior UK astronomers had had their way and SERC had voted for an 8-metre partnership with the Spanish on La Palma. Optimistically, it might have led to a credible European northern observatory, a long-held dream of Sánchez that was repeatedly snubbed by a succession of ESO directors general. The Italians were developing a 3.6-metre national telescope, the Telescopio Nazionale Galileo. Although they had also preferred to locate this on Maunakea, they eventually completed it on La Palma in 1988 following disappointing discussions with the University of Hawaii. Conceivably, other non-ESO countries might have followed, but this did not happen, as ESO continued to accrete new members and expanded its facilities in Chile (chapter 9).

Pessimistically, however, a UK-Spanish partnership might have taken several more years to reach a detailed plan for construction and shared instrumentation. This had been Wolfendale's major concern in 1990. The 8-metre project on La Palma might not have been ready for SERC's funding approval in 1994 and, given the worsening financial situation at the time, might even have been abandoned as unaffordable or undeliverable. The credibility of UK ground-based astronomy would then have been seriously dented in government circles, with enormous consequences for the future. Without question, the decision to go with the United States and Canada set the direction for UK ground-based astronomy for decades to come.

To his enormous credit, Sánchez did not give up when abandoned by the United Kingdom. Spain eventually raised the funds for its own national facility, the Gran Telescopio Canarias (GranTeCan, or GTC), an ambitious 10.4-metre segmented-mirror telescope modelled on the Keck telescopes (chapter 7). The construction of the GTC took 7 years and cost 130 million euros (plate 32). Minor partners include Mexico and the University of Florida. Whereas I attended the ground-breaking and dedication ceremonies of the twin Gemini telescopes in 1995 and 1999 (plate 30, *right*), GTC did not begin science operations until 2009, more than a decade after most of the international suite of 8-metre telescopes were operating.

Today, thanks largely to Sánchez's persistence, La Palma remains an important astronomical observatory, one that survived the UK decision to contribute instead to the Gemini Observatory, as well as the expansion of ESO member states. Inevitably, both the United Kingdom and Spain also joined ESO in 2002 and 2006, respectively. Fifteen years after the controversial SERC decision in 1990, I met Paco Sánchez in Mexico. He always used to sign his personal letters to me with the phrase "Un abrazo" (A hug). As we met, he expressed no animosity at all for my part in the UK decision to go elsewhere and gave me that hug.

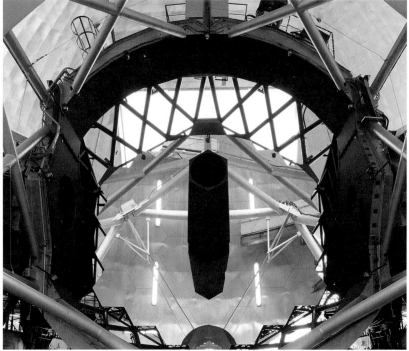

PLATE 32. The Spanish 10.4-metre Gran Telescopio Canarias was completed in 2007 and is the largest optical telescope on La Palma. Modelled on the 10-metre Keck telescopes in Hawaii, its primary mirror is composed of 36 hexagonal segments.

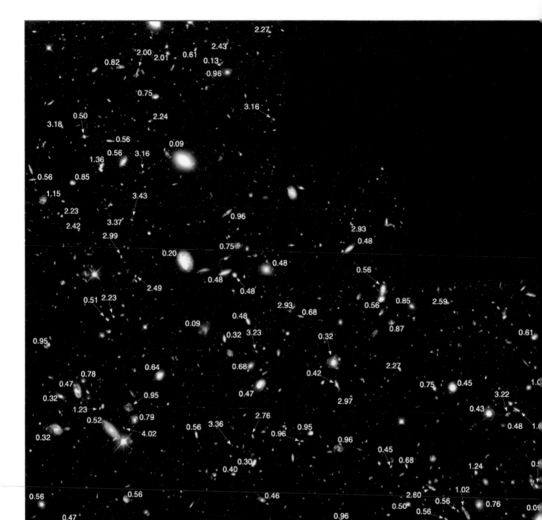

PLATE 33. Colour image of the Hubble Deep Field released in January 1996. The L-shaped format reflects the fact that one of the four detectors in the WFPC2 has a finer image scale. This remarkably deep image, taken at the initiative of STScI director Bob Williams, captured the public's interest in the distant universe and inspired many astronomers to measure spectroscopic redshifts for the brighter galaxies (marked) over the next few years.

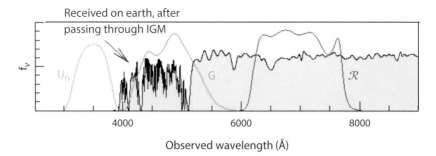

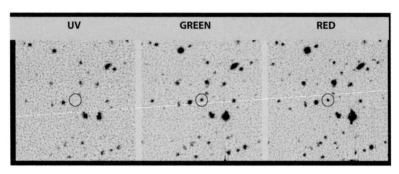

PLATE 34. Illustration of the "dropout" technique for locating candidate high-redshift galaxies prior to confirmatory spectroscopy. (*Top panel*) How the energy spectrum (yellow) of a distant galaxy is absorbed by neutral hydrogen in the galaxy and along the line of sight in the intergalactic medium shortward of a known rest wavelength. In this particular example, the galaxy is at a redshift of $z \sim 3$ and the rest wavelength is stretched by the expansion of the universe to 4000 Ångstroms in the near ultraviolet. (*Lower panel*) It is possible to locate such a distant galaxy from others in the field of view by imaging in three filters. The $z \sim 3$ galaxy disappears ("drops out") in the UV filter, whereas the others, at lower redshift, do not.

PLATE 35. A night of observations beckons. Sunset on the summit of Maunakea (alt. 4200 metres), on the Big Island of Hawaii. The Subaru 8-metre telescope is at the left, next to the twin Keck telescopes. The adjacent island of Maui can be seen in the direction of the setting sun.

PLATE 36. (*Right*) The Keck I telescope pointing horizontally during the afternoon, revealing its 10-metre primary mirror composed of 36 hexagonal segments. (*Left*) Down at Waimea, Caltech students Dan Stark and Matt Schenker (foreground) take part in a night-time observing session. Matt controls the instrument and Dan analyses the incoming data for discoveries.

PLATE 37. The French connection. (*Left*) At the influential 1989 Toulouse gravitational lensing meeting. The author is on the right next to Geneviève Soucail, and Yannick Mellier is on the far left. Both were pioneers of the early days of lensing. (*Right*) On the summit of La Palma with Jean-Paul Kneib (next to the author on the far right) and Roser Pelló (with jacket on shoulder, fourth from the left).

PLATE 38. Hubble Space Telescope image of the cluster Abell 2218. Our earlier version of this image convincingly demonstrated the power of gravitational lensing as a tool for studying distant galaxies. The large, bright (orange) galaxies in the cluster are at a redshift $z = 0.17$, and the highly distorted, mostly blue, galaxies are background sources up to redshifts of $z \sim 2.5$.

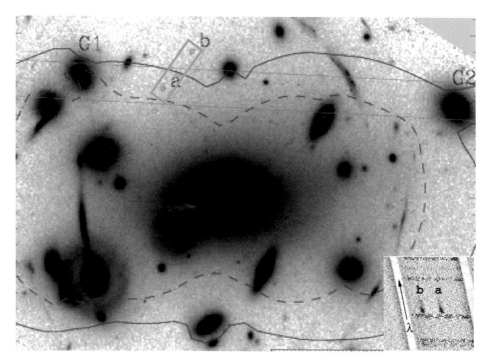

PLATE 39. The critical curve method. The solid red curve represents the predicted location of very highly magnified sources at a redshift of $z \sim 6$, behind the cluster Abell 2218 (shown in this negative image). The dashed red line refers to the location of the critical curve for sources at much lower redshift. Kneib, graduate student Mike Santos, and I searched the area contained within the blue rectangle and located image a. Once we ascertained the redshift of this image, Kneib's geometrical model for the cluster predicted a counter-image b. On a second run we searched the area within the green rectangle. The inset at bottom right shows the detection of hydrogen emission at identical redshifts $z = 5.576$ for both a and b.

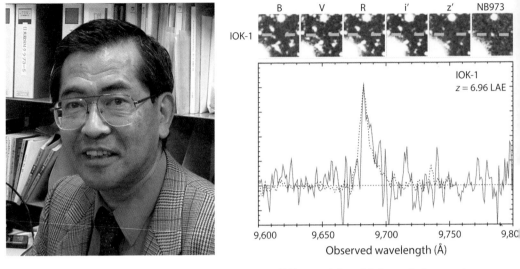

PLATE 40. A Japanese team led by Masanori Iye (*left*) secured the redshift record of $z = 6.96$ in 2006 with IOK-1. (*Right*) The top panel shows how the source is visible only in a narrow band filter NB973 (in between the blue pointers), already suggesting it is at a redshift $z \sim 7$. Following this up with a heroic 15-hour spectrum (bottom panel) on the Subaru Telescope, his team confirmed a hydrogen emission line whose observed wavelength gives the precise redshift.

PLATE 41. A collection of Plumian Professors at Cambridge on the occasion of Fred Hoyle's 80th birthday in 1995. From left: Martin Rees (1973–1991), Fred Hoyle (1958–1972), myself (1993–1999), and Bertha Jeffreys, whose late husband, Harold Jeffreys, held the chair from 1946 to 1958.

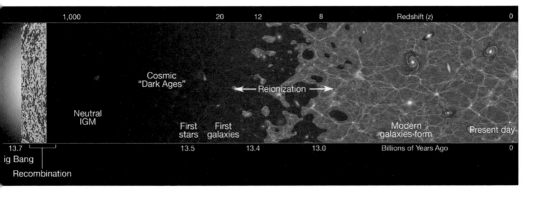

Cosmic
"Dark Ages"

◄— Reionization —►

Neutral
IGM

First   First
stars   galaxies

Modern
galaxies form   Present day

13.7   13.5   13.4   13.0   Billions of Years Ago   0

ig Bang

Recombination

PLATE 42. Illustration of the reionisation era with time running from left to right. As the universe expands it cools, positively charged protons capture negatively charged electrons, and hydrogen atoms are eventually formed (recombination). The dark ages end when the first stars and galaxies form and their energetic radiation detaches electrons, creating patches of ionised gas that grow in size and eventually coalesce until the entire intergalactic space is fully ionised.

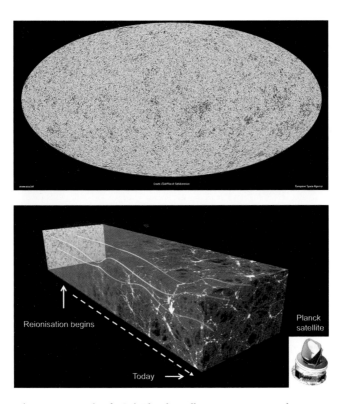

PLATE 43. The important role of ESA's Planck satellite in constraining when reionisation occurred. (*Top*) Map of the temperature fluctuations in the microwave background seen 380,000 years after the Big Bang. (*Bottom*) As the radiation from this early era passes through space it is polarised by electrons from the time when reionisation began to the present era.

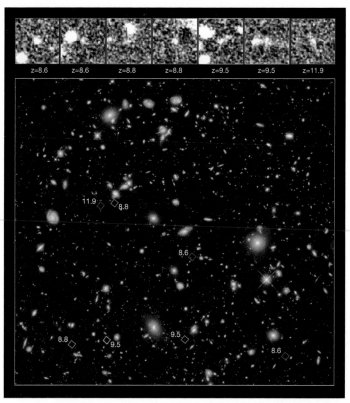

| z=8.6 | z=8.6 | z=8.8 | z=8.8 | z=9.5 | z=9.5 | z=11.9 |

PLATE 44. The Hubble Ultra-Deep Field (HUDF12) campaign. (*Top*) A group photo taken (modestly as ever) by Brant Robertson at a planning meeting at the Royal Observatory, Edinburgh, in 2012. On the author's left is Jim Dunlop, and to his left is Ross McLure. (*Bottom*) The resulting near-infrared image remains the deepest ever taken with Hubble, and the seven newly found galaxies beyond redshift 8.6 (including the enigmatic UDFj-39546284 at redshift 11.9) are marked in the panels above with their predicted redshifts.

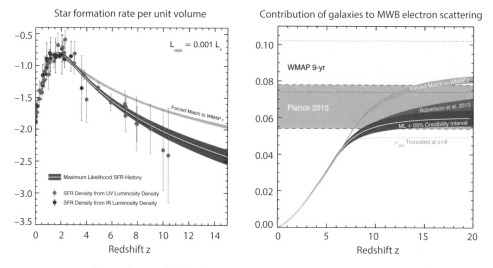

Star formation rate per unit volume

Contribution of galaxies to MWB electron scattering

PLATE 45. Reconciling results from the HUDF12 campaign with electron scattering measures from the microwave background (MWB). (*Left*) Number of star-forming galaxies per unit volume as a function of redshift (blue points). The red band shows a fit to the HST data. (*Right*) The red lines now indicate how far back in redshift this population must exist to reproduce the electron scattering seen by Planck. Unsuccessful attempts to fit the earlier WMAP results are shown in orange.

PLATE 46. Cerro Tololo (left) and Cerro Morado (right) in the Atacama Desert, 500 kilometres north of Santiago. Tololo is the home of the Inter-American Observatory operated by the Association of Universities for Research in Astronomy (AURA). Cerro Morado was briefly considered as a site for an international partnership between ESO and the United States.

PLATE 47. (*Upper panels*) Cerro Paranal from a distance and closer up at sunset. The four unit telescope (UT) enclosures are open in preparation for a night's observations. (*Lower left*) UT4 (Yepun). Each telescope has two elevated Nasmyth platforms for large instruments and a Cassegrain focus centred below. (*Lower right*) No expense spared! The ESO residence has a small swimming pool and a pleasant garden. The apparatus suspended from the glass ceiling opens fully at night to form a light-tight curtain to avoid disturbing night-time observations over a kilometre away.

PLATE 48. The beginning of ESO's "next big thing," the Extremely Large Telescope, as seen in late 2015. The summit of Cerro Armazones was flattened with explosives in 2014. Red bollards outline the area which will contain the foundations of the 39-metre aperture telescope.

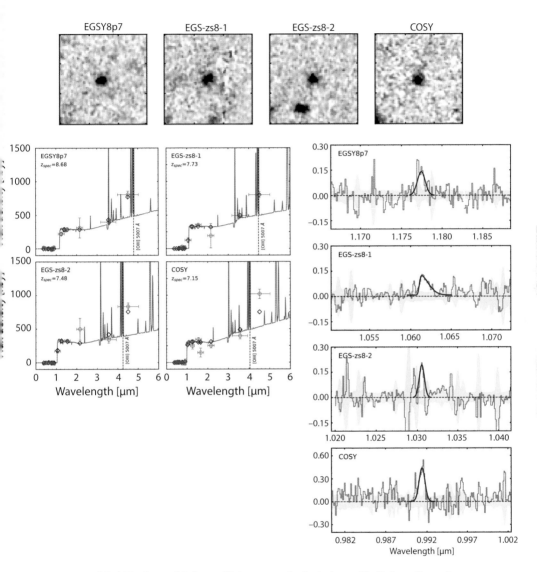

PLATE 49. (*Top*) The four redshift $z > 7$ "Spitzer-excess" galaxies located by Roberts-Borsani. (*Lower left*) Hubble and Spitzer measures of their brightness as a function of infrared wavelength (in microns, μm). Red points represent the observations, which in each case reveal a sharp jump in the Spitzer brightness at 4.5 microns. Blue lines and black diamonds represent model spectral fits, indicating the Spitzer excess coincides, at these high redshifts, with extraordinarily intense oxygen emission (vertical dashed lines). (*Lower right*) Lyman-alpha emission (red curve) at the spectroscopic redshifts indicated on the left.

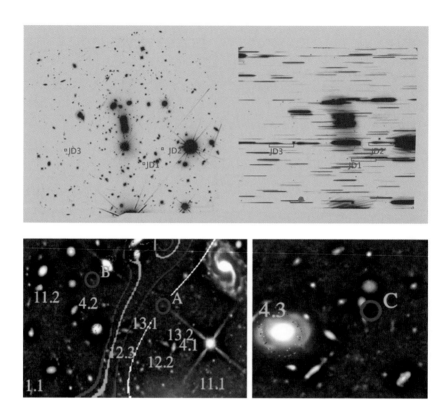

PLATE 50. The challenge of confirming the redshifts of two $z \sim 10$ gravitationally lensed sources. (*Top*) Nor Pirzkal's attempt to secure the redshift of the gravitationally lensed source MACS0647-JD, whose limited colour information indicated a redshift $z \sim 10.7$. The left panel shows three images of the same source in the field of the foreground cluster and the right panel illustrates the challenge of extracting the grism spectra amongst the numerous overlapping signals. (*Bottom*) Three images (A, B, C) for the gravitationally lensed source A2744-JD1. Coloured lines show the location of the critical line of very high magnification for a source at redshift 1 (white), 2 (blue), 3.5 (green) and 10 (red). Zitrin argued that the geometric arrangement of the three images means the source must be at redshift greater than $z \sim 4$.

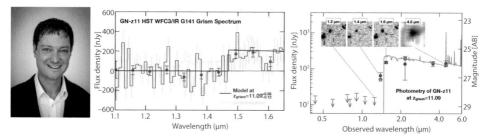

PLATE 51. Pascal Oesch (*left*) presented a redshift $z = 11.1$ candidate on the basis of an HST grism spectrum (*centre*) of the source GN-z11, previously known as GN-z10-1. The sharp drop in signal (red line) just below 1.5 microns (μm) is due to redshifted hydrogen absorption. The Hubble imaging (*right*) shows a detection at 1.4 microns, suggesting such absorption is at a lower redshift of $z = 10.2$–10.4. Oesch and colleagues claimed both data sets are consistent with the higher redshift.

PLATE 52. Sunset at Cerro Paranal with Guido Roberts-Borsani (centre) and Nicolas Laporte (right) prior to determining the redshift and age of MACS1149-JD1 with ESO's Very Large Telescope.

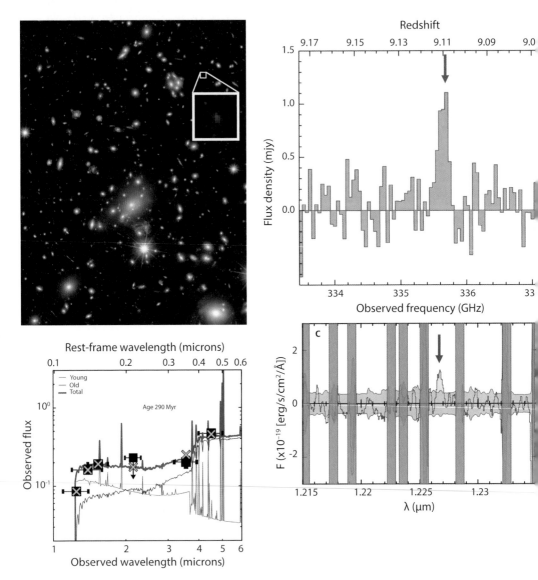

PLATE 53. A first indication of when cosmic dawn occurred based upon MACS1149-JD1. (*Top left*) Hubble image of the foreground cluster MACS1149 + 2223 with a zoom on JD1. (*Right panels*) Securing the same redshift $z = 9.11$ using the Atacama Large Millimeter Array (top) and ESO's X-shooter spectrograph (bottom). (*Lower left*) The brightness of JD1 with wavelength, demonstrating that the Spitzer excess at 4.5 microns (µm) is from a Balmer break and not intense oxygen emission (red spike). The strength of this Balmer break indicates an age of 290 million years, leading to a first estimate of cosmic dawn at about 250 million years after the Big Bang. Black and blue curves represent, respectively, mature and younger stellar components in JD1.

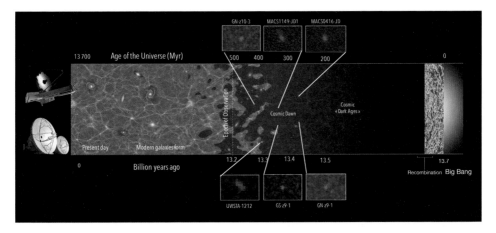

PLATE 54. Updating when cosmic dawn occurred. The results of age-dating six redshift $z \sim 9$ galaxies from the observational campaign led by Nicolas Laporte. The epoch of formation ranges from 250 to 350 million years after the Big Bang, corresponding to the redshift range $z \sim 13$ to 16, an epoch that can be directly probed with the James Webb Space Telescope. Myr, million years.

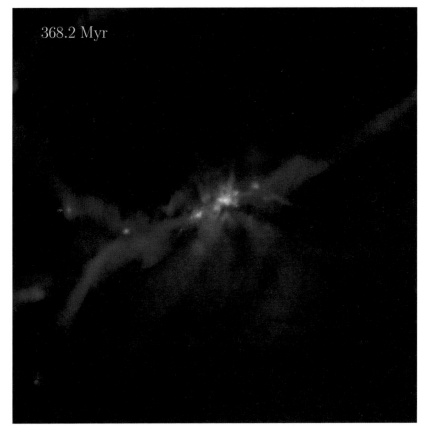

PLATE 55. Witnessing cosmic dawn. A snapshot from a numerical simulation in a virtual universe similar to own. The image shows galaxy similar to those studied by our research group seen close to its birth, when the universe was just over 350 million years old redshift $z \sim 13$). Purple regions display the filamentary distribution of gas, composed mostly of hydrogen. White regions represent arlight and the yellow regions depict energetic radiation from the most massive stars, which is capable of ionising the surrounding ydrogen gas. As massive stars rapidly reach the end of their lifetime they erupt in violent supernova explosions, which expel the surrounding gas, enabling the escape of this energetic radiation. Galaxies such as the one shown continually accrete material from nearby maller systems and quickly assemble to form the more-substantial galaxies observed by the Hubble Space Telescope at later times.

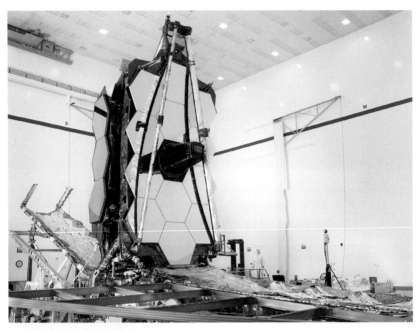

PLATE 56. (*Top*) The James Webb Space Telescope (JWST) under final assembly and testing at the Northrop Grumman facility at El Segundo, Los Angeles, in late 2019. To give a sense of its size, a technician can be seen to the right of the base of the mirror. (*Bottom*) View of JWST on Christmas day 2021, 15 seconds after separation from the upper stage of the *Ariane 5* launch vehicle over east Africa at an altitude of 1413 kilometres and moving at 10 kilometres per second. This is likely our last direct view of this impressive space telescope.

6

A Golden Era

Synergies with Hubble Space Telescope

In 1977 I cut out a full-page advertisement that appeared in the *Financial Times*, a London- based daily newspaper that focuses on business and economic affairs. Placed by the Lockheed Corporation, an American aerospace company, it showed an illustration of NASA's Space Telescope (later named the Hubble Space Telescope, HST, after the observational cosmologist Edwin Hubble) in orbit. The headline caption proudly announced: "In 1983 man may see to the edge of the universe." I pasted this cutting on my office wall in Durham, and each time there was a delay, as so often happens with space missions, I amended the expected date when we might achieve this long sought-after vision of the early universe (figure 6.1). My foresight in keeping this advert over the next 13 years was noted by many who visited my office, but, for those of us in our early careers at the time, my action represented our eagerness to exploit the new telescope and our frustration at the delays, since we knew it would transform our understanding of the distant universe.

Establishing an optical observatory in space was a dream that began in the 1940s. Not only would such a facility be above the clouds and capable of taking data without the restriction of cloud cover that affects ground-based telescopes, but it would also be free from the blurring or "seeing" caused by the turbulent layers of the Earth's atmosphere. The latter advantage would lead to sharper images revealing morphological details in galaxies, such as spiral arms, to much greater distances. As we

FIGURE 6.1. The tattered remnant of the Lockheed advertisement in the 1977 *Financial Times*, with my record of the seemingly ever-receding launch date of the Hubble Space Telescope.

saw in chapter 4, the night sky is a major limitation to studies of very faint galaxies. From a ground-based observatory, this background signal from the sky is dominated by airglow—the radiation produced by the effect of sunlight on atoms in the Earth's upper atmosphere. Above the Earth's atmosphere, the background signal away from any celestial target can be nearly 10 times darker.[1] For this reason, a telescope in space

1. It may be surprising to some to learn that the sky is not completely dark in space. Most of the background signal in outer space is the so-called zodiacal light, a diffuse glow that straddles

can be as sensitive for studies of faint galaxies as a much larger one on the ground. And, as discussed earlier, a space observatory can also study regions of the electromagnetic spectrum, such as the ultraviolet and infrared, that are absorbed by the Earth's atmosphere.

The downside of a space observatory is the staggering financial outlay. Whereas 4-metre telescopes completed in the 1970s, such as the AAT, cost a few million dollars at the time, what became the 2.4-metre Hubble Space Telescope cost close to $2 billion, and this is before adding the cost of servicing missions using NASA's space shuttle. I failed to keep track of the rising cost of HST during the late 1970s and 1980s, but it was clearly a political issue. Enthusiastic supporters often claimed HST would be the "eighth wonder of the world," to which a sceptical congressman once replied, "It ought to be at that price." Contemplating HST in the late 1970s was the first time I realised that astronomy was moving into "big science." Whereas the AAT served primarily UK and Australian astronomers, and Palomar Observatory belonged to Caltech and Carnegie Observatories, HST would be truly a one-off—a unique facility that had no competitor. Although funded by the US and European space agencies, it would become a global facility for all astronomers for decades. The remarkable history of the Hubble Space Telescope has been told in great detail by other authors, and I won't repeat it here.[2]

Of the many astronomers who devoted their careers to ensure its eventual success, my personal choice is two individuals from Princeton whom I met several times (figure 6.2). Lyman Spitzer (1914–1997) was one of them. He was an American theoretical physicist and astronomer who pioneered studies of the interstellar medium—the material between the stars. In the 1930s, it was realised that stars are powered by nuclear fusion. Theorists calculated that although our sun could burn

---

the ecliptic, the sun's path through the sky. It is caused by the scattering of sunlight by interplanetary dust particles.

2. A well-researched book that tells the detailed story prior to launch in 1990 is *The Space Telescope: A Study of NASA, Science, Technology and Politics* by Robert W. Smith (Cambridge University Press 1989).

FIGURE 6.2. Two influential astrophysicists who ensured the scientific success of the Hubble Space Telescope, Lyman Spitzer (*left*) and John Bahcall (*right*). Spitzer promoted the scientific gains of an orbiting observatory as early as 1946 and was an earlier pioneer in space astronomy. Both raised community support for HST and lobbied tirelessly for its funding, rescuing it from cancellation on more than one occasion. Bahcall was later influential in maximising the efficiency of the observatory.

for billions of years, more-massive stars would last only a few hundred million years. Spitzer realised that stars are continually being created from the "interstellar" material in between them. This cycle of stellar birth from interstellar gas that, for the most massive stars, leads to supernova explosions that in turn replenish the gas with processed material is a fundamental feature of cosmic evolution.

To investigate the physical properties of this interstellar material required access to ultraviolet observations above the Earth's atmosphere. Spitzer wrote an influential paper in 1946 entitled "Astronomical Advantages of an Extraterrestrial Observatory," and, after pioneering such ultraviolet space missions in his early career, he later became the champion promoting the cause for a large space telescope (LST). He chaired a seemingly endless number of future planning committees concerned with garnering community support, which ultimately led to a NASA study of a 3-metre LST in 1973. Moreover, Spitzer later convinced the US Congress to fund the project, and it eventually became the 2.4-metre HST. The fact that LST could also stand for the

"Lyman Spitzer Telescope" was not lost on many astronomers, who recognised his leadership of the project. In many ways, Spitzer could be regarded as a modern-day equivalent of George Ellery Hale (chapter 3).

I met Spitzer for the first time in the late 1970s when I was working on the scattering and polarisation of starlight by interstellar dust particles. We were both attending a conference at Cambridge in England, walking along the street together to the Institute of Astronomy. Here was the director of the famous Princeton department and arguably one of the most influential American astronomers talking to a young postdoc he'd never met before. Yet, interested in my work, he offered to have lunch with me. When I searched for him at noon he was besieged by senior Cambridge astronomers inviting him for lunch, but he turned all these down to honour his promise to me. His charming modesty and willingness to spare his precious time with me impressed me greatly.

The other astronomer crucial to the success of HST was John Bahcall (1934–2005), a professor at the Institute of Advanced Study at Princeton. Bahcall became a member of a Science Working Group for the Space Telescope in 1973 and, with Spitzer, the most effective proponent arguing for its funding. Given how successful it eventually became, it might be surprising to read that the Space Telescope was a controversial project in the 1970s. Many astronomers feared its significant cost would preclude continued investment in ground-based telescopes, and those scientists studying the solar system were competing within NASA for missions to the outer planets. But more fundamentally, the ever-increasing cost due to the inevitable technical delays associated with its construction led to fierce debates within the US government committee responsible for NASA appropriations. Several times the telescope was threatened with cancellation. Bahcall excelled at promoting the project within government circles and raising support through vigorous lobbying from the entire US astronomical community and relevant industrial partners. At one House Appropriations subcommittee hearing, a congressman commented that he "didn't know there were so many astronomers in the United States, frankly, until the last six months. They all have typewriters . . . and they can find their way to the Post Office."[3] Possibly Bahcall's greatest service to HST was his forceful advocacy for resurrecting the Space Shuttle servicing missions after the *Columbia* shuttle disaster in 2003. This enabled a final servicing mission in 2009 that has extended the scientific lifetime of HST by at least a decade.

Despite all of the above, Bahcall was fundamentally a respected theoretical astrophysicist. He made exacting calculations that predicted the arrival rate of solar neutrinos (elementary particles produced in the fusion that makes the sun shine). His predictions were discrepant with observations, leading to his formulation of a "solar neutrino problem." Bahcall worked on the topic, often dismissed by physicists and astronomers as an unimportant detail, for three decades. It was eventually resolved by the discovery that neutrinos can oscillate from one type (or "flavour") to another. While Bahcall correctly predicted the solar production rate, the early experiments were sensitive to only one neutrino flavour. However, during their passage from the sun to Earth, neutrinos change from one flavour to another. The resolution of this long-standing problem led to the 2002 Nobel Prize in Physics being awarded to two experimentalists but not to Bahcall, a disappointment to many.

As well as being extraordinarily influential in political circles and one of the world's greatest astrophysicists, John Bahcall's personal style operated at two very different levels. On the one hand, he was a caring mentor and genuinely interested in supporting young people. As a regular visitor to Princeton, he and his wife Neta (also an astrophysicist) took me under their wing and supported me many times when I was contemplating a move to another institution. On the other hand, John represented a formidable presence at the Institute of Advanced Study in Princeton. He ran a famous "Tuesday lunch," which could be a fearful event for any visitor. Following a morning colloquium, lunch was arranged at a large U-shaped table. There, John would introduce the visiting speaker to a large audience of Princeton scientists (including several members of the US National Academy) and begin by asking

3. *The Space Telescope*, Robert W. Smith (Cambridge University Press 1989), p180.

the speaker to summarise the earlier talk in a single sentence. He would then switch gears and, out of the blue, ask for a description of another recent scientific accomplishment from the speaker. The unpredictability of the occasion, and the possibility of making a blunder in the presence of so many dignitaries, could be either terrifying or exhilarating depending on one's self-confidence. Regardless of which, the event was certainly memorable!

By the early 1980s, the *Financial Times* article on my office wall indicated HST would be launched in 1986. The next issue concerned how it should be operated. For a space facility as complex as HST, it seemed obvious to NASA managers that their agency should run the observatory. Given the enormous government investment, they argued oversight was required from an experienced national organisation. On the other hand, HST was the first major space facility whose purpose was solely to enable astronomers to do research. Spitzer and others argued the scientific operation should therefore be managed to serve its users, the astronomers. Although the tension between NASA and scientists for control of space missions goes back to the 1960s, it now resurfaced in a major way for HST. Through a subgroup led by Arthur Code, an experienced space scientist, John Bahcall played a key role in pushing for an independent HST science institute under the charge of a consortium of universities. The astronomical community was wary of NASA control, and it took further persistence to avoid locating this independent science institute at a NASA centre. After a bidding process in 1980, it was agreed the newly formed Space Telescope Science Institute (STScI) would be placed on the Homewood Campus of the Johns Hopkins University in Baltimore. Its inaugural director was the experienced X-ray astronomer Riccardo Giacconi (1931–2018). Many were surprised that a Princeton bid had not been selected, with John Bahcall possibly appointed as director.

The European Space Agency (ESA) is a 15% partner in HST, and young European astronomers like me were being exhorted to familiarise themselves with the scientific instruments being developed for HST so as to be ready when the first call for observing proposals was made. In 1984 ESA set up its own centre, the Space Telescope European

Coordinating Facility (ST-ECF) near Munich. There were numerous workshops dedicated to preparing proposals, and the atmosphere was one of considerable excitement. In the spring of 1985, a year before the expected launch, I was invited to STScI in Baltimore for a 3-month visit. This was a great opportunity to learn all about HST. The institute was buzzing with activity—for example, in developing software tools to enable the community to prepare observing proposals based on the sensitivities of the various scientific instruments on board HST. It was my first extended visit to the United States. My children, aged 7 and 8, attended local schools in Baltimore, and we enjoyed exploring New York and Washington, DC, as well as making an excursion to Toronto for a conference.

As the launch date approached, one of the major questions facing senior astronomers at STScI was how to ensure the best science was done with HST. Most ground-based observatories issue a call for observing proposals once or twice a year. These calls are usually unrestricted, in the sense that there are no preordained scientific topics: the community has free rein to submit any proposal of its choice, and those with the most compelling scientific arguments are approved for observing time. Although HST was expected to last at least a decade, given that any technical failures could be rectified by a space shuttle visit, there still was a strong belief by those advising the STScI director and senior staff that, by virtue of its unique capabilities, there were some key science questions that "HST simply *must* address." These "key projects" included ones to address fundamental science questions that had, over the past decade, been repeatedly emphasised to justify HST. But who should frame these key projects, and how would the community participate in executing them? Through its advisory committee, STScI set up seven scientific working groups to identify such key projects. Once agreed, the list would be announced to the community, and each project would be assigned a provisional amount of observing time on HST. Different groups of astronomers would then submit proposals and compete to undertake each key project.

I was selected to be a member of the Science Working Group for Surveys, whose charge was to consider how to use the so-called Wide Field Camera (WFC) on board HST to undertake untargeted large-scale imaging surveys. WFC was the brainchild of Jim Westphal (chapter 3), its principal investigator, then employed at the Jet Propulsion Laboratory in Pasadena, and Jim Gunn at Caltech was an influential team member.

I was the only European serving on the committee and in distinguished company. Other members included my galaxy-count rivals Richard Kron and Tony Tyson; the famous cosmologist Jim Peebles, whom I had first met almost a decade earlier (see chapter 4); Alan Dressler, initially a competitor but later a collaborator in studies of distant clusters of galaxies (see the epilogue); and John Bahcall. Our 1985 report is fascinating to read today after 30 years of successful HST operations. Although our recommended surveys were eventually undertaken, the scientific motivations we gave in our report are puzzling in hindsight. I have found this to be the case for many ambitious projects, including large ground-based telescopes. The scientific case originally made to justify their funding doesn't tally that well with the eventual highlights after a decade of operation. In nearly all cases, the facilities significantly outperformed the earlier predictions and discovered the unforeseen.

Our working group recommended two types of surveys in the early years of HST operations. The first was a Medium Deep Survey (MDS) to be conducted in "parallel mode." This mode exploits the fact that the various scientific instruments (camera, spectrographs, etc.) on board HST can simultaneously access different portions of the focal plane, a gain in efficiency that Bahcall had proposed some years earlier. For example, while HST points to a specific target with one of its spectrographs, the imaging camera can take a picture of an adjacent area of sky. Our working group envisaged a parallel WFC imaging survey of 50 fields, each taken with a short exposure, exploring the counts of faint stars and galaxies, as well as searches for unusual objects ranging from possible star clusters that lie in between galaxies to comets beyond the orbit of Neptune. The second campaign that we recommended was an Ultra Deep Survey which would pull out all the stops and undertake much deeper exposures targeting only a handful of random fields on the sky that were free from bright stars. This would count yet fainter stars and galaxies and was justified with rather nebulous statements about learning more about the evolution of galaxies in order to constrain cosmology.

There was a surprising emphasis in our report on counting faint stars to constrain models of the outskirts of the Milky Way. This was probably due to the influence of John Bahcall on the working group. He had produced the most definitive models of the structure of our Galaxy and, given his heroic efforts to realise HST, it was natural that we should respect his push for this science topic. Nonetheless, we already knew from earlier studies of deep photographic plates (chapter 4) that there would be far fewer stars than galaxies at the faintness limits likely to be reached by HST, and the small field of view of WFC would not lead to statistically reliable counts. What was much more surprising, however, was the continued hope stated in our report that studies of faint galaxies in the Ultra Deep Survey could somehow be used to constrain the cosmological parameters. The report concluded, "The idea is to use all of the obtainable information in a synergistic way to determine evolutionary effects. Given evolutionary corrections, the test for (cosmic deceleration) could include a comparison of the coadded frames, and the redshift distribution, with simulations."[4] Even a decade after the aborted efforts to use distant galaxies to constrain cosmology using the Hale 200-inch at Palomar, it seemed there was still a desire to keep trying to pursue Hubble's programme.

Certainly the most important message in our report was the emphasis on the need for ground-based spectroscopy to secure the redshifts of the faint galaxies in both surveys, in order to understand how galaxies evolve in their morphological forms. HST cannot compete in securing galaxy redshifts with large ground-based telescopes equipped with multi-object spectrographs. This "ground-space synergy" was to become a major advance in studies of faint galaxies as detailed below.

On January 28, 1986, the first proposals were being submitted to STScI for observing time on HST, due for launch in October that year. Prior to the internet, these were typed on paper forms and mailed to Baltimore. After lunch in Durham, we took our package of proposals to the university mail room, where they would be collected by a courier

4. Report of the Space Telescope Scientific Working Group for Surveys, March 21, 1985, unpublished report held at the Space Telescope Science Institute, Baltimore, MD, quote at p11.

and sent to the United States. A few hours later Tom Shanks appeared in my office and silently motioned with his hands the rise and precipitous fall of a rocket. That very morning in Florida the Space Shuttle *Challenger* had exploded, killing seven astronauts. I remember, with horror, the insensitive television coverage, which included the real-time reactions of relatives of the astronauts invited to witness the launch from the VIP area at Cape Canaveral.

For the next 7 years, the HST mission seemed truly jinxed. Having recently spent 3 months in Baltimore, my heart went out to those 200 or so scientists and engineers at STScI who had worked so hard to prepare for HST operations and now faced an uncertain future. The space shuttle fleet was grounded for almost 3 years, but HST was eventually launched on the Space Shuttle *Discovery* on April 24, 1990, a delay of over 4 years. Initially considered a success, it soon emerged the telescope had a major problem. Although stellar images taken with HST had a sharp core, there was an extended halo that could not be eliminated by refocusing. The news that HST could not be brought into focus was initially suppressed by NASA but, inevitably, it was eventually revealed to HST's Science Working Group, which comprised senior members of the US astronomical community. These included Jim Westphal, principal investigator of the optical camera, and Sandra Faber, a prominent Californian observational astronomer (introduced in more detail in chapter 8). Together with Chris Burrows, an optical expert at STScI, the trio conclusively demonstrated that HST suffered from a fundamental optical defect referred to as "spherical aberration." The primary mirror, polished to exquisite precision by Perkin-Elmer, had the wrong shape, owing to the incorrect assembly of a component used to test the mirror. At that fateful Science Working Group meeting, Sandra Faber announced, "Our scientific program is fully compromised—devastated."[5]

Like many, I had been given approved observing time on HST, and the images I eventually received looked weird. Distant galaxies showed

---

5. A cynical and somewhat abrasive account of this period by a former STScI scientist involved in the HST project is given in *The Hubble Wars: Astrophysics Meets Astropolitics in the Two Billion Dollar Struggle over the Hubble Space Telescope* by Eric J. Chaisson (Harper Collins 1994), p181.

some sharp features that would not be seen from a ground-based telescope, but they sat in a milky halo of light that rendered precision measurements impossible. As one astronomer commented at the time: "The telescope is working well enough to show us what we're missing."

A few mathematically minded astronomers claimed it ought to be possible to "undo" the deleterious effect of spherical aberration since, they argued, the "point spread function" (the spatial response of an optical system to a point of light) of the misshaped mirror was well understood. Two radio astronomers at Cambridge had a private company that was using such deconvolution techniques to assist the British police in revealing car license plates from blurred images. I spent a frustrating week at the ST-ECF in Munich trying out their various numerical tricks on powerful computers, but the improvements were marginal and affected measurements of the brightness of the galaxies (figure 6.3).

As a result of the misshaped primary mirror, HST immediately became an embarrassment, widely ridiculed in the press and referred to variously as "Hubble trouble," "myopic telescope," or "techno-turkey." In Washington, DC, a new verb was coined, "to hubble," meaning to screw up in a major unfocused way.[6] Fortunately, from images taken with HST and studies of the test piece retained by Perkin-Elmer, it was possible to precisely calculate the error in the shape of the primary mirror and design a set of correcting optics. These optics were installed during the first servicing mission with the Space Shuttle *Endeavor* in December 1993. Some existing instruments, such as the WFC, were replaced with ones that incorporated the correcting optics (e.g., the Wide Field Planetary Camera 2, WFPC2), while others had the correcting optics retrofitted. At the American Astronomical Society Meeting in January 1994, spectacular fully corrected images were shown to the astronomical community to cheers of joy. John Wilford Noble, a well-respected reporter for the *New York Times*, said: "Science news has to involve big bang, big bucks, big screw-up or big come-back—and with Hubble you've got them all!"[7]

6. *The Hubble Wars*, Eric J. Chaisson (Harper Collins 1994), pp192–95.

7. Ibid., p367.

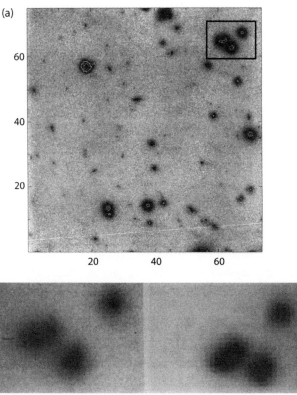

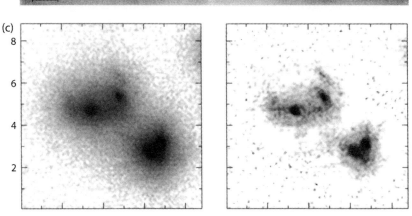

FIGURE 6.3. (*Top*) HST WFC negative image of a redshift $z = 0.32$ cluster taken prior to the 1993 servicing mission. Each galaxy is surrounded by an extended halo of light. (*Centre*) HST image of three galaxies in the black box in the top panel (left) compared with a ground-based equivalent (right); HST clearly reveals finer detail but much light is spread into the halos. (*Bottom*) Deconvolving the HST image using the known spherical aberration reduces the halos somewhat (right), but the overall improvement in resolution is marginal and quantitative measures of brightness are difficult to make.

The first comprehensive HST survey of faint galaxies was the MDS, following the recommendations of the Key Projects Working Group for Surveys 1985 report. I was a member of the winning team, led by Richard Griffiths, a fellow Welshman based at Johns Hopkins University in Baltimore. He was an experienced X-ray astronomer and member of the WFPC2 science team. Although familiar with the route to science from telescope observations (i.e., writing the proposal, gathering and analysing the data, and submitting an article for publication), I soon learned that working with HST introduced several additional complications. Firstly, there was a general belief that to get large amounts of HST observing time, large teams were essential. Our MDS team had 15 members, of which 11 were in the United States, thus strong leadership and coordination were essential. To accomplish this management necessitated regular "telecons" to plan and review progress; these usually interrupted my family evenings given that our team included members spread over eight time zones. Next, a successful HST proposal led to associated grant funding from NASA, which paid not only the stipends of graduate students and postdocs for assistance with data analysis but also the salaries of senior team members. I did not appreciate until then that most US universities pay the salaries of their professors for only part of the year; they are expected to raise their "summer salaries" from such grants. It gradually dawned on me that, for many US astronomers, securing a summer salary seemed to be a more important reason to apply for HST time than even pursuing the science. As principal investigator, Richard Griffiths was in charge of distributing the large NASA grant among the various US co-investigators (non-US members were ineligible for funding). There were often fierce accusations among the team that Griffiths was hoarding the lion's share of funds for his own use. As I could not receive any funds myself, I was appointed to arbitrate on such matters. This unfortunate responsibility led to several US members besieging me for support via phone calls ahead of team meetings where the distribution of funds was to be discussed.

The MDS survey was approved in the first annual call for proposals (referred to as Cycle 1) in late 1988 prior to launch, and hence before the mirror problems were revealed. As with most Cycle 1 approved programmes, because of the misshaped primary mirror, the relevant observations were deferred. Although some data were taken in 1992–1993, most of the high-quality data were taken after the first servicing mission, between 1994 and 1996.

In "parallel mode," the fields that HST imaged were decided by "primary" pointings undertaken by programmes using other instruments. Consequently, there were no redshifts available for any of the galaxies we observed. In essence, the MDS represented a return to simply counting galaxies and measuring their colours as Kron, Tyson, and my colleagues and I had pioneered over 10 years earlier. The main gain, of course, was the improved angular resolution. It enabled us to classify the faint galaxies by their morphologies broadly into three classes—spirals, ellipticals, and irregulars. By 1993 I had moved from Durham to Cambridge with my postdoc Karl Glazebrook (chapter 4). We arrived to find a welcoming group excited about exploiting the new Hubble images. It included David Schade, Harry Ferguson (see later in this chapter), and Rebecca Elson, a talented and courageous woman who tragically died from lymphoma in Cambridge aged only 39.

With a Canadian postdoc at Cambridge, Roberto "Bob" Abraham, we began counting hundreds of galaxies across the various MDS images and classifying them according to their visual morphology. My mundane responsibility was to examine each galaxy on my computer screen and classify it as a "spiral," "elliptical," or "irregular." The latter category was inevitably a catch-all for pretty well anything that didn't look smooth and well formed. Although we did not have individual spectroscopic redshifts for the MDS sample, we could predict from our AAT spectroscopic surveys that most of these galaxies would, statistically, be seen out to redshifts of about $z \sim 1$, corresponding to a lookback time of 7 billion years. Whereas the number of faint spirals and ellipticals seemed to be in accordance with expectations, given their local abundances, an unusually high proportion of the faintest MDS galaxies were irregular in form (figure 6.4). A deeper image of one field taken by a group at the University of Arizona confirmed this result. Together with the bluer colours of the fainter irregulars, this supported our earlier

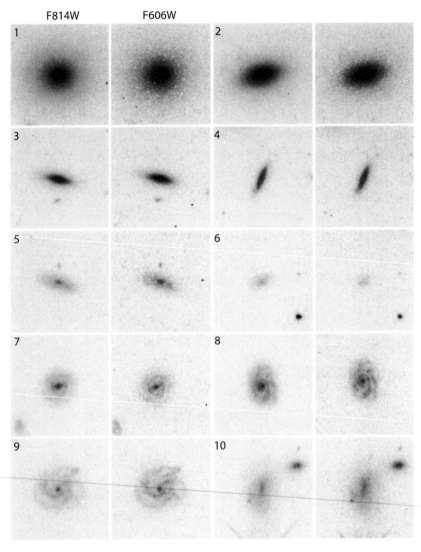

F814W   F606W

FIGURE 6.4. Morphological classifications from Medium Deep Survey images taken after the servicing mission. This 1994 analysis was based on nine MDS fields and visual classifications were undertaken by the late John Huchra and me. Two colour images of 10 regular galaxies (*this page*) are contrasted with those of 20 peculiar and irregular systems (*facing page*), which are more numerous at faint limits. Not all irregulars were observed through two colour filters. Although no redshifts were available at the time, from surveys to similar brightness limits, the galaxies are expected to lie in the redshift range 0 to 1.

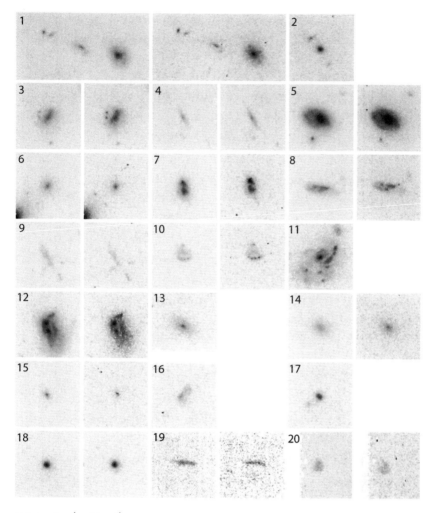

FIGURE 6.4. (*contiinued*)

hypothesis (chapter 5) that the universe was more active in star-forming irregulars at earlier times, but a key question was what happened to those peculiar-looking galaxies. Did they somehow transform into more normal-looking galaxies? As with the photographic count studies, we were witnessing another manifestation of evolution but were unclear of the physical processes involved.

Working with Glazebrook and Abraham in Cambridge was one of the most productive periods of my time in the United Kingdom. I was

newly arrived at the Institute of Astronomy, which was originally established by Fred Hoyle in 1967 as a theoretical institute. The main building
(later named after Sir Fred), was purposely designed to be conducive
to scientific interactions. Morning coffee at 11 a.m. was a lively event
attended by all, and in summer we could stroll to the lush green surroundings in warm sunshine and discuss the latest HST results. Bob
Abraham, in particular, had an almost childlike enthusiasm for astronomy that was infectious. He was keen to automate the process of classifying the Hubble images (and put me out of business!), and so we
explored the use of his technique, which involved a classification based
on two parameters—central concentration and asymmetry. The degree
to which a galaxy image had a central concentration of light exploited
Hubble's superlative resolution to identify regular ellipticals or the central bulge of a spiral galaxy. The asymmetry index, obtained by considering quantitively whether one side of the image was an accurate reflec-
tion of the other, could be used to locate irregulars. We tested the
reliability of these simple parameters on a sample based on my visual
classifications. Abraham's automated classification technique was found
to be very effective.

Meanwhile at STScI, Riccardo Giacconi, the inaugural director,
moved to Europe to become director of the European Southern Observatory (ESO; chapter 9) and was replaced in 1993 by Robert "Bob" Williams, previously a director of the US-funded Cerro Tololo observatory
in Chile (figure 6.5). I found Bob much more approachable than Giacconi, and we got on very well, particularly when I found out that his
ancestors came from a small town in North Wales barely 18 miles from
my birthplace. As STScI director, Williams had a mission to promote
more widespread use of HST data by the global community. Much of
the early HST data was proprietary, reserved for exclusive use either by
those who proposed the observations or those scientists on the instrument teams, who were rewarded handsomely with guaranteed observing time. Although all HST data did eventually emerge into a publicly
available archive, many of the exciting results had already been extracted
by those with proprietary access. In an imaginative move, Williams decided to release the data from a significant allocation of the time he had

FIGURE 6.5. Robert "Bob" Williams, director of the Space Telescope Science Institute from 1993 to 1998, initiated the Hubble Deep Field (HDF) campaign that established the tradition of the rapid release of publicly available and fully reduced HST datasets for use by the global community. The HDF campaign transformed both the general public's and the astronomical community's interest in exploring the distant universe.

at his discretion as director for immediate public release. In 1995, he gathered together a bunch of astronomers, including myself, to discuss for what scientific purpose this discretionary time should be used.[8] According to my notes, "Those present will remember a rather rambling discussion with . . . much disagreement on details." Williams had a vision for an extremely long exposure in a single unremarkable patch of sky, a so-called Hubble Deep Field (HDF). Since this would be taken with the WFPC2 camera, which has a field of view only a tenth of the diameter of the full moon, it would be like a deep pencil beam penetrating into the distant universe. The visiting astronomers disagreed on

8. The fascinating story of how Bob William's campaign transformed faint galaxy studies with HST is described in detail in *Hubble Deep Field and the Distant Universe* by R. Williams (Institute of Physics 2018), which is also available free online as an e-book.

whether it was better to spread such a large amount of observing time over several areas of sky, or perhaps to point at a target already known to be of astrophysical interest. As I concluded soon after, "We hardly prescribed HDF at that meeting."

Disregarding the disagreement among his chosen panel of advisors, Williams pressed on with planning his HDF campaign and harnessed a team of enthusiastic young postdoctoral researchers at STScI, including Harry Ferguson and Mark Dickinson, to select the optimum colour filters and relative exposure times. Exposure time on HST is often reckoned in terms of orbits. Although the orbital period of the spacecraft is around 96 minutes, depending on the sky position of the target, not all of that time is usually available for scientific exposures. This is because the target can be occulted by the Earth during part of the orbit. However, certain regions of the sky, called "continuous viewing zones," are not occulted, and so for maximum observing efficiency the HDF field was chosen to be in one such northern position. Dickinson had already demonstrated the remarkable power of HST in deep imaging a field containing radio galaxy 3C324 at a redshift $z = 1.2$. The total exposure time was 32 orbits, much longer than the deepest MDS images of 5 orbits. The eventual HDF campaign involved a total of 150 orbits taken over a period of 10 days in December 1995.

I was surprised to read in Williams's account that both John Bahcall and Lyman Spitzer, the two most distinguished long-standing supporters of HST, were concerned there might not be many galaxies beyond the limits seen in Dickinson's 32-orbit image at redshift $z \sim 1$. Although this now seems a curious objection, given HST has probed galaxies out to redshifts beyond $z \sim 9$, it must be remembered that in 1995 astronomers were very uncertain when galaxies first formed. Bahcall and Spitzer argued that, if only a handful of more distant galaxies were found, the HDF campaign would be viewed as a colossal waste of observing time that would bring Hubble back into disrepute in government circles because of the recent $1 billion cost incurred by taxpayers to repair Hubble's flawed optics. Williams deserves much credit for maintaining the courage to press on and explore the most basic question of all: "What's out there?"

The spectacular HDF image did not disappoint, and put all concerns to rest (plate 33). It revealed over 3000 faint galaxies, ranging from exquisitely detailed spirals and ellipticals via a plethora of ever fainter and smaller irregulars to ultimately tiny dots of light that seemed to beckon the viewer into a deep abyss. As I was travelling when the data were released in 1996, I first saw a colour image of the HDF, constructed from the four WFPC2 filters, on the front page of a daily newspaper while sitting on an aeroplane. I was gobsmacked![9]

Back in Cambridge, Bob Abraham and I already knew that the HDF image and its associated data would be publicly released on January 15, 1996. In order to be the first to analyse and publish results from what was widely regarded as the most exciting step forward in exploring the faint universe, Abraham took the lead in drafting a scientific article *before* receiving the data, leaving blanks for the results and figures! When the data were released, Sidney van den Bergh (a veteran Canadian astronomer and one of Abraham's heroes) and I worked overtime to morphologically classify over 3000 faint galaxies in the image and Abraham ran his asymmetry-concentration computer code on each galaxy. The resulting article, the first exploiting the HDF, was posted on the internet on January 31, barely 2 weeks after the image was released. We extended the validity of the evolutionary trends seen in the MDS to a new limit around 16 times fainter. Although the number counts continued to increase, the bulk of the faintest galaxies were irregular in form, not regular ellipticals and spirals (figure 6.6). Presumably they were very distant sources that had yet to establish their regular forms. A couple of days after posting the article on the internet, I got an email from Bob Williams at STScI saying that our article had been discussed at their weekly science discussion meeting in a packed auditorium at Baltimore where it was standing-room only!

It was obvious to many when the HDF was being planned, and even more so when the spectacular image was revealed, that spectroscopic redshifts were key to understanding the details of this morphological evolution. Could we honestly be sure these irregulars and tiny smudge-like

9. British slang for "astounded."

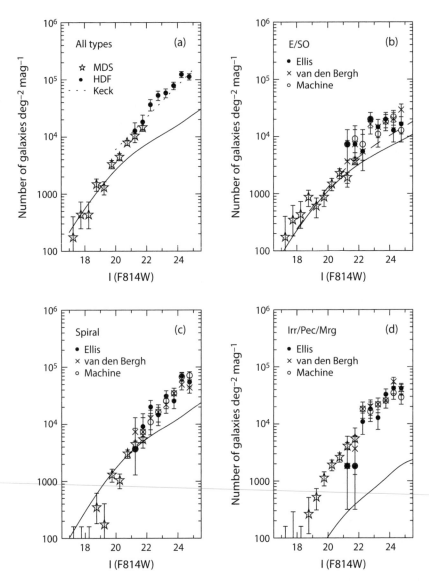

FIGURE 6.6. Galaxy counts per unit sky area by morphological class versus apparent magnitude in the HDF (fainter sources have larger magnitudes). The *top-left panel* indicates the total counts of all galaxy types and demonstrates the much increased depth compared to the earlier Medium Deep Survey. The other three panels show the counts for ellipticals (*top right*), spirals (*bottom left*), and irregular/peculiar or merging systems (*bottom right*), according to visual classifications by Van den Bergh and myself and Abraham's automated asymmetry-concentration code (labelled Machine). The solid curve indicates a no-evolution prediction based on the present-day abundance of galaxies of different types. The bulk of the evolution arises from an increased abundance and luminosity of irregular and peculiar galaxies.

galaxies were the most distant galaxies in the picture? After all, we had already found that the excess galaxies seen in the shallower AAT surveys led by Tom Broadhurst were not necessarily more distant but the result of complex luminosity-dependent evolution. Since the location of the tiny HDF field was chosen on the basis of HST observing efficiency, ground-based observers had not studied this field before. Nonetheless, once the HDF coordinates were known, those with access to the newly completed 10-metre Keck telescope on Hawaii began to hammer away at securing spectra. Within only 18 months of the release of the data, thanks to huge efforts by Mark Dickinson and Len Cowie (chapter 5) and others, there were 125 redshifts secured in this small field (plate 33). The majority lay within the redshift range $0 < z < 1$, further confirming that the excess counts to the deepest limits yet achieved arose from more actively star-forming galaxies of modest luminosity.

An interesting feature of the HDF image, which I first noticed while examining the newspaper photograph on the aeroplane flight, was that there was still quite a lot of blank sky in between all the galaxies (see plate 33), despite the extraordinarily long exposure time. Could it be, I wondered, that we were peering through all of cosmic history back to a time when there were no galaxies at all? Tony Tyson, the photographic galaxy counter, working with Princeton graduate student Puragra "Raja" Guhathakurta, had already suggested a clever trick for constraining how far into the past the faintest galaxies may spread. They pointed out that hydrogen gas, both in galaxies and in clouds along the line of sight in the intergalactic medium (the space in between galaxies), strongly absorbs radiation in the far-ultraviolet spectral region. This absorption occurs at a very specific wavelength, owing to the energy levels of the hydrogen atom, and renders a galaxy invisible at shorter wavelengths. Since the wavelength of this absorption "cut-off" is redshifted by the cosmic expansion into the bluest portions of the optical spectral range by a redshift $z \sim 3$, galaxies beyond this redshift would disappear when examined in a carefully chosen blue filter. From deep CCD images in various filters, Guhathakurta and colleagues found none of their galaxies disappeared in their bluest filters and therefore argued that all must lie below a redshift of $z \sim 3$. However, since quasars were already known to redshifts of $z \sim 4.7$ at this time, it wasn't clear whether this conclusion precluded yet fainter galaxies at higher redshifts.

A major breakthrough occurred when a Caltech graduate student, Charles "Chuck" Steidel, began using the Hale 200-inch at Palomar to study the regions around high-redshift quasars. By deep imaging high-redshift quasars with a carefully chosen set of optical filters, he began locating what he termed "dropouts"—galaxies present in two filters that disappeared in the shortest wavelength filter—consistent with their being associated with the quasar and the cut-off of hydrogen absorption at this redshift (plate 34). The redshifts of these dropouts were to some extent validated, since they lay adjacent to a much brighter quasar whose redshift was known. This encouraged Steidel to be more ambitious and extend the campaign to other fields without the reassuring presence of a quasar of known redshift. He later spectroscopically confirmed that these dropout candidates were indeed above a redshift $z \sim 3$ using the Keck telescope. When the HDF data appeared, Dickinson, Steidel, and others used the same technique with the WFPC2 filters to locate 69 galaxies consistent with being in the redshift range $z \sim 3$ to $z \sim 3.5$. In this way it was, for the first time, possible to address the abundance of star-forming galaxies beyond a redshift $z \sim 3$, corresponding to a lookback time of nearly 11.5 billion years, when the universe was only 17% of its present age.

A clear picture began to emerge from this dramatic observational progress. The early galaxy counts and redshift surveys demonstrated a major increase in the total amount of star formation seen in galaxies back to a redshift $z \sim 1$; this was the main explanation for the long-standing puzzle of the excess number counts and the abundance of immature irregular galaxies. Nonetheless, if the small HDF was representative of the faint universe, the collective star formation implied in the 69 dropouts beyond a redshift $z \sim 3$ was less than that observed at a redshift 1. In other words, the HDF image probed sufficiently far back in time to witness the galaxy population in decline—as I had suspected looking at the blank spots on my newspaper.

Two years after the HDF data release, STScI held an international conference in Baltimore to celebrate the research done with this single remarkable data set. As with most astronomical conferences, there was insufficient time for all attendees to have the opportunity to speak, so many brought posters to highlight their work. In addition to 3 full days of science talks to an audience of over 150 participants, there were over 50 posters displaying the results of teams from all over the world who had undertaken wide-ranging analyses of this unique Hubble image. One had to congratulate Bob Williams; through his director's initiative, he had invigorated worldwide interest in the distant universe. On the other hand, I suddenly felt newly challenged. My research field had completely transformed. Not only had Chuck Steidel and his colleagues opened up the high-redshift universe to observations far beyond the redshift $z \sim 1$ limit I had achieved at the AAT and WHT over a decade of building multi-object spectrographs, but suddenly there were hundreds of young researchers interested in observations of the early universe. This was a far cry from the cosy territory of galaxy counting on photographic plates occupied by only Tyson, Kron, and myself!

7

The Mighty Keck Telescopes

Lifting the Curtain

The summit of Maunakea, a (reputedly) dormant volcano on the Big Island of Hawaii, is arguably the best location worldwide for astronomical observations. The telescopes perched at 4200 metres altitude sit well above the typical cloud inversion layer, where the humidity is low, and the oceanic isolation of this enormous mountain (which extends to a similar depth under the sea) provides a tranquil airflow that ensures excellent seeing conditions. The reduced mass of air above the summit compared with that at sea level has earned Maunakea the oft-quoted distinction of being "halfway to space." As a result, in the 1990s, when technology and funds permitted the construction of a new generation of telescopes with mirrors of diameter 8 to 10 metres, astronomical institutions from the United States and Japan secured permits to build large facilities on Maunakea.

The site was first recognised as valuable for astronomical research by Gerard Kuiper (1905–1973), a planetary scientist at the University of Chicago, and in the late 1960s the University of Hawaii, whose primary astronomy department resides in a leafy district of Honolulu on the island of Oahu, erected the first major telescope with a 2.2-metre-diameter mirror near the summit. Noting environmental concerns even at that time, the state of Hawaii created a Science Reserve on the summit established under a lease from the state's Board of Land and Natural Resources, with day-to-day management allocated to the university.

Astronomers recognise the privilege of observing on this remarkable mountain, whose summit is considered sacred in the Hawaiian religion. Indeed, by ancient law, access was restricted to high-ranking nobility. Native Hawaiians have a strong historical connection with the sky; they arrived from Polynesia over 1000 years ago using remarkable skills of celestial navigation. Today, all scientific publications based on data taken with the telescopes on Maunakea acknowledge the very significant cultural role and reverence that the summit has always had within the indigenous Hawaiian community.

I first visited Maunakea in the early 1980s to use the 3.8-metre UK Infrared Telescope (UKIRT), which was completed in 1979. At that time, observers stayed in primitive accommodations at a mid-level facility called Hale Pohaku—necessary preparation for acclimatising us to working at the summit, where the oxygen content is roughly 60% of that at sea level. A medical examination was mandatory for all observers, and altitude sickness, mostly manifesting as headaches, was a frequent complaint on the first night of observations. On rare occasions, astronomers had to descend rapidly to avoid risk of a pulmonary or cerebral oedema.

Despite these drawbacks, I fell in love with Maunakea. There was a real sense of working at the frontier—from manoeuvring a four-wheel-drive vehicle along the precarious, unsurfaced road to the summit to watching the sun rise with the shadow of this majestic mountain cast over the Pacific Ocean and the other Hawaiian islands clearly in view (plate 35). At that time, the telescopes were used mostly for infrared observations, which took advantage of the dry atmosphere at high altitude. Today, astronomers don't like being categorised by the wavelengths of their observations, but in those days "infrared astronomers" were a wild bunch (then mostly male and Americans), sporting lumberjack-style checked shirts, several days' growth of beard, and a swashbuckling attitude. They were surprisingly welcoming to an inexperienced Brit. Not only did I enjoy their camaraderie but I began to realise that American astronomers could be fun to be with.

Although many 2- to 4-metre telescopes were built on Maunakea in the 1970s and 1980s, the Keck Observatory was the first 8- to 10-metre telescope to arrive, in 1993. Designed and built through a partnership between the privately funded Caltech in Pasadena, California, and the state-funded campuses of the University of California, the project was the brainchild of Jerry Nelson (1944–2017), an astronomer and particle physicist who devoted his career to innovations in telescope design. Early on, Nelson had realised that to fabricate a mirror twice the diameter of the Hale 200-inch (5-metre) at Palomar would be impractical for many reasons, not least with respect to transporting it and preventing the mirror from deforming under its own weight. He had the bright idea of a *segmented mirror*, composed in this case of 36 hexagonal mirrors, each independently supported and positioned mechanically under computer guidance, so that together they could maintain the precise shape of a single curved surface (see also chapter 2).

I first heard about Nelson's idea in the late 1980s when I was chairing the Large Telescope Panel (LTP) for the UK Science and Engineering Research Council (chapter 5). We were charged with making the case for the UK astronomical community's construction of an 8- to 10-metre telescope. A consultant on our committee, David Brown, was a former employee of the telescope-manufacturing company Grubb Parsons of Newcastle-upon-Tyne, which had built pretty much all of the United Kingdom's telescopes up to that point. Brown had vast experience in telescope making and was exceedingly sceptical about Jerry Nelson's segmented mirror plans, as were many others in the business. Certainly Nelson hit many hurdles along the way—for example, in polishing each segment to the required shape—but he eventually succeeded through sheer persistence, ingenuity, and extensive prototyping. Built at a cost of approximately $70 million, the 10-metre telescope was funded by the Keck Foundation, established in 1954 by the oil entrepreneur William M. Keck, and began operations as planned in 1993, under the joint auspices of Caltech and the University of California. The most astonishing aspect of the Keck story is that, during the construction of the first telescope, the foundation, on its own initiative, decided to fund a second telescope as well. It came online in 1996.

I didn't meet Nelson until I moved to California in 1999, but when I did I soon realised he was a formidable figure. He seemed amiable enough, always smiling, and with a good sense of humour. However, although never directly aggressive, he was, as might be expected of such an unconventional thinker, a natural contrarian who seemed to relish playing devil's advocate. On more than one occasion, I witnessed him politely interrogating less-experienced astronomers and technical staff to the point where he capably demolished their impractical plans. With his almost godlike status within the community, nobody dared challenge his authority. Nelson seemed remarkably healthy and fit—he once told me he enjoyed swimming across the majestic Waipio Bay when in Hawaii—and I was one of many colleagues who was shocked and dismayed when he suffered a massive stroke in 2011. It is the measure of the man that, as his health permitted, he continued to work at his office in Santa Cruz, California, almost up to the time of his passing, which sadly took place in 2017.

Back in the United Kingdom, the imminent arrival of not just one but *two* 10-metre telescopes solely for the use of California astronomers threatened to send the global balance of large-telescope access back to the days of Palomar's dominance, which had lasted from 1950 to the mid-1970s. Although my LTP was having promising discussions about building two 8-metre telescopes in a partnership with Canada and the United States, which would eventually become the Gemini Observatory (chapter 5), and Europe had ambitious plans concerning the construction of *four* 8-metre telescopes, these projects were years behind the twin Kecks.

Each of the Keck telescopes has nearly six times the light-gathering power of the 4.2-metre William Herschel Telescope on La Palma in the Canary Islands, which was the most powerful facility I had access to in the mid-1990s. Recording the signal of a distant galaxy with the same level of fidelity as an astronomer using one of the Keck telescopes, an observer less fortunately situated would have to expose their target *at least* six times longer. "At least" because weather conditions and instrumental problems inevitably affect routine observations. And, fundamentally, time on any telescope is a precious commodity, in great demand by the observational community. I might get three or four nights every 6 months on the Herschel telescope, and a California astronomer might expect a similar allocation on the Keck. For this reason it is best to think of what can be accomplished in that amount of time. In short, to keep up with those Keck observers, I'd have to get 18 or 24 nights on the Herschel every few months, and such generosity of telescope access was simply out of the question. I recall a colleague at Cambridge posting an early Keck spectrum of a distant quasar on the departmental notice-board with a gloomy caption that the United Kingdom was back in the second division of world astronomy.

My term as director of the Institute of Astronomy in Cambridge was coming to an end in 1999, and the equivalent position at the Institute for Astronomy at the University of Hawaii was vacant. The previous long-serving incumbent, the lively Australian Don Hall, had stepped down, and his chosen successor, Frank Shu, a distinguished Berkeley astrophysicist, eventually turned down the offer. Colleagues in Hawaii invited me to apply for the vacancy, and I flew to Honolulu in late 1998 for an interview. This was a huge adventure. Unlike many of my UK colleagues, I had never spent any extended time outside my home country because, only a few years after my PhD, my wife Barbara and I already had two children. Nonetheless, Hawaii was appealing not just because of its exotic location but also because these large telescopes on Maunakea provided a unique opportunity for discovery. In exchange for managing the Maunakea site, Don Hall had successfully secured a 10%–15% share of observing time for the University of Hawaii *on each of the summit's telescopes!* As a potential director overseeing the management of all these observatories, I envisaged, perhaps naively, orchestrating their coordination to maximise their overall scientific output. In short, I was ready for a challenge.

Professors never like to advertise to their departmental colleagues that they have itchy feet. Imagine the consequences if they fail to secure a particular job elsewhere. "You're evidently not seriously devoted to our department, so why should we listen to your opinions?" would be a typical remark. As a result, senior faculty shuffles in academia are usually kept confidential but often become the subject of gossip. Exactly how my interest in Hawaii leaked out, I'm unsure. But on one dark November afternoon in Cambridge, I got a phone call from Chuck Steidel at Caltech, asking if I'd be interested in a position in Pasadena. If so, would I be passing through soon so I could meet a few people? It was obvious from the timing of this "casual inquiry" that word had somehow got out that I would be transiting California on my way to Hawaii. The long and short of this interesting period of my life is that I ended up accepting a position at Caltech in spring of 1999. My wife and I moved there in September to set things up, and I officially became a Caltech professor in December.

In the spring of 2000, I got my first real taste of observing at the Keck telescopes. Conducting observations remotely via the internet from the US mainland was not then practical, so Californian astronomers would fly to Kona on the west side of the Big Island and stay at a residence located right next to the Keck headquarters in the sleepy town of Waimea, once renowned for its cattle ranches. After all my nights struggling with altitude sickness and time spent jolting along the rock-strewn road to and from the summit, observing on a dedicated internet link from the Keck headquarters in Waimea seemed like luxury. Nonetheless, I was nervous about the prospect of using one of the world's largest twin telescopes for the first time. Taking into account the salaries of over 100 operational staff, every second of observing time was worth over one dollar. Screwing up was not an option!

My experience of hundreds of nights of observing over two decades mostly in Australia and the Canary Islands did not prepare me for the major surprise of using the majestic Kecks. The first step in locating a faint galaxy at the edge of the universe is what we call "target acquisition." Normally this is done using a television camera to locate a nearby "reference star" whose sky coordinates relative to the faint galaxy are known from pre-existing pictures. The astronomer then shifts the telescope by the known angular distance to the (normally undetectable) faint galaxy, whose light then enters a spectrograph, which measures the galaxy's redshift and other properties. This process, known as a "blind offset," requires precision and great faith. With smaller telescopes, I often struggled to identify the correct reference star amongst a number of faint smudges on the camera. Selecting the wrong one would mean offsetting to the incorrect position and might lead to a wasted 1- or 2-hour exposure that altogether missed the much fainter target galaxy. At the Keck, however, there was no such uncertainty. I was blown away to see that the television image contained an abundance of beautifully point-like stars and even extended galaxies. Using the powerful 10-meter aperture was like lifting a curtain and seeing a clear view of the heavens for the first time. This was a formative experience for me. Not only did I immediately realise the astonishing scientific promise of this remarkable instrument, but it inspired me to be more ambitious. Although astronomers often propose observational programmes that are too difficult, somehow with the Keck anything seemed possible (plate 36).

At Caltech, all observing proposals are assessed and judged by a time allocation committee (TAC), which meets in person and ranks maybe 100 proposals from faculty and postdocs using scientific merit, previous scientific output, and technical feasibility as criteria. Keck time would be scheduled based on the scientific ranking of the proposals, taking into account the visibility of targets in the night sky. If you wanted time when the moon was below the horizon (when the sky is at its darkest) in the month of your choice, your proposal had better have a high rank.

Arriving at Caltech as a foreigner I had been impressed by how welcoming everyone was. Pretty soon I was invited to give talks all over the country. But I quickly learned that in the wider US astronomical community Caltech astronomers were regarded with some suspicion, maybe jealousy, and only occasionally grudging respect. It was a commonly held view that Caltech allocated enormous amounts of observing time on its coveted telescopes to its own faculty for programmes that were not seriously reviewed, and that certain Caltech astronomers, over years of plentiful observing, had compiled vast amounts of data that still hadn't been analysed. Meanwhile, astronomers in less-privileged US institutions either had no access to Keck at all or struggled to get only a few nights a year on far less powerful facilities. At the time of my arrival, Caltech enjoyed an enviable 40% of the observing time on the two Keck telescopes, divided amongst perhaps only 15 faculty and 20 postdocs. Meanwhile, all the various campuses of the University of California, comprising more than 100 active astronomers, had to compete for a similar amount of time. So it is true that Caltech astronomers were in a privileged position. However, during my subsequent years as director of the Caltech Optical Observatories, a position responsible for the technical development and scientific oversight of Caltech's share of Keck, I was a regular chair or member of the TAC, and I always found the standard of observing proposals to be very high. Occasionally, of course, some astronomers acted as spoilt children, arguing that they had a "right" to telescope access. Most, however, agreed that writing a convincing proposal and demonstrating successful outcomes with earlier time allocations was a fundamental requirement. The simple fact is that the scientific productivity of Keck has always been outstanding compared with peer observatories. Caltech's privileged access enabled extraordinary science and was very rarely wasted.

Let me now turn to the first project I undertook given my new-found access to the Keck telescopes, which was based on exploiting the magnification of distant galaxies by foreground structures. Gravitational lensing—the bending of light rays by massive celestial bodies, resulting in brighter, magnified, and in some cases multiple images of distant cosmic objects—has a fascinating history. As early as 1704, a possible germ of this idea can be found in an appendix entry of Isaac Newton's *Opticks*. Newton wrote, "Do not Bodies act upon Light at a distance, and by their action bend its Rays; and is not this action strongest at the least distance?"[1] The nature of this query is the subject of some dispute among historians; however, Newton's speculation is intriguing to say the least.

Nearly a century later, calculations of the deflection of starlight by the sun were made independently, in 1784 and 1801. However, those calculations were based on Newton's theory of light, where it was assumed light particles undertook trajectories around the solar limb with a varying velocity, similar to the path of a cannonball in the Earth's gravitational field. Developed early in the twentieth century, Albert Einstein's special theory of relativity does not permit the velocity of light to vary, and a further contribution to the deflection arises from his general theory—which holds that gravity arises because of the curvature of

1. *Opticks*, 4th edition, I. Newton (1730; repr. Dover 1952), book 3, part 1, p339.

space. The combination of both of these effects doubled the traditional estimates, leading to a deflection of 1.75 seconds of arc; this is a small effect equivalent to the angle subtended by a person's height at a distance of over 200 kilometres.

Although Einstein urged several observers to detect this effect, the first successful attempt occurred during a now-famous solar eclipse in May 1919. A campaign organised by the British Astronomer Royal, Frank Dyson, undertook measurements of stellar positions at the time of the eclipse in both Sobral, Brazil, and Príncipe, an equatorial island off the west coast of Africa. The distinguished astronomer Arthur Eddington (1882–1944) led the expedition to Príncipe, and his confirmation of Einstein's prediction later that year catapulted the young theorist to international fame. In something of a pilgrimage, I visited both Príncipe and Sobral in the late 2000s. Príncipe is no longer a Portuguese colony but part of the independent state of São Tomé and Príncipe. Recently it has expanded its tourist industry significantly, including building a new hotel close to the spot where Eddington made his measurements. Sobral is a major town in the Brazilian state of Ceara and has a museum dedicated to the historic eclipse.

Paradoxically, neither Einstein nor Eddington believed gravitational lensing would have any practical utility other than verifying relativity. Surely, they argued, the likelihood of two stars being aligned (other than the case where one star is the sun) would be minuscule. Einstein undertook some of the relevant calculations but concluded "of course, there is no hope of observing this phenomenon directly."[2] The first person to challenge this established wisdom and to claim gravitational lensing might be a valuable astronomical tool was Fritz Zwicky (1898–1974), a prickly, notably eccentric, but undoubtedly visionary Caltech professor, who suggested this idea in 1937. His insight was to perceive that clusters of galaxies had a significant cross-section on the sky and so could act as giant cosmic lenses. By deflecting and magnifying the light from background sources, they could, in principle, enable astronomers to see

2. Lens-like Action of a Star by the Deviation of Light in the Gravitational Field, A. Einstein, *Science*, vol 84, no. 2188 (1936), p506.

extremely faint objects at great distances. Since Einstein was a regular visitor to Caltech in the 1930s, surely he and Zwicky discussed the idea, but on this the record appears to be silent. Although they were colleagues, I have not been able to find any correspondence between them on the practical aspects of gravitational lensing in either the Caltech archives or in Caltech's Einstein Papers Project.

Although several theorists worked on the mathematics of gravitational lensing in the 1960s, its potential as a tool for observational astronomy remained dormant until 1979 when two apparently identical quasars were found by an Anglo-American team using the 2.1-metre telescope at Kitt Peak National Observatory in Arizona. These "twin quasars" are separated by only 6 arcseconds, and it was subsequently shown that both images were in fact the result of light from a single quasar being deflected by a galaxy midway between the quasar and the Earth. More dramatic evidence of gravitational lensing emerged in the 1980s when astronomers discovered highly distorted "arc-like" images of galaxies in the vicinity of well-known clusters of galaxies. In 1987, spectroscopy by a French team led by Geneviève Soucail at Toulouse Observatory verified that one of these stretched images represented the magnified light of a more distant galaxy, magnificently confirming Zwicky's prediction in 1937. Overnight, Soucail became a media star. I remember her modestly explaining how she was embarrassed to pose under a photograph of Einstein for an endless number of journalists.

I was a professor at Durham University at the time, but I already had strong connections with the now famous group of French observers in Toulouse. I was inspired by their discovery and the enthusiasm evident in the youthful team of Soucail, Yannick Mellier, Roser Pelló, and their leader Bernard Fort (plate 37). In 1989, a memorable conference in Toulouse dedicated to the prospects of gravitational lensing marked a turning point in the subject as well as for my own career. I became convinced that harnessing the magnifying power of galaxy clusters would lead to great progress in probing the early universe. The enthusiasm at the conference was infectious, but I remember Roger Blandford, a distinguished British astrophysical theorist then at Caltech who gave the conference summary, warning that observers should be careful not to over-interpret their data. He joked that some astronomers seemed to perceive every close pair or elongated image on the sky as arising from a gravitational lens.

In 1993 I moved to Cambridge University and organised and informally led a collaboration of European scientists to further research into gravitational lensing. The team was made up of astronomers at the universities of Cambridge, Leiden, Munich, and Toulouse. The European Union gave each institute funds for a postdoctoral assistant. I had the good fortune to hire Jean-Paul Kneib, a French graduate from the Toulouse group that had led the way in exploiting clusters as gravitational lenses. He was in Chile at the time, working there for the European Southern Observatory, which operates telescopes in the Atacama Desert (chapter 9), and I interviewed him for the Cambridge position in a coffee shop in Santiago. A keen mountaineer, during his time there he made an impressive ascent of Aconcagua, at 6900 metres the highest peak in South America.

The partnership with Kneib and one of my former Durham students, Ian Smail, was very productive. In 1996 we wrote a proposal to secure a Hubble Space Telescope image of the cluster Abell 2218. The Toulouse group had already located several possible distorted arcs in this cluster using Maunakea's ground-based Canada-France-Hawaii Telescope. Orbiting above the distorting interference of Earth's atmosphere, Hubble is able to obtain images much sharper than those even from the largest ground-based observatories. Kneib, Smail, and I had a hunch that our Abell 2218 image might be impressive, so I wrote to the director of the Space Telescope Science Institute, Bob Williams (chapter 6), to ask if I could get immediate access to the image once it was taken. I was giving a public lecture to the British Association for the Advancement of Science on the topic of gravitational lensing in a few days' time and I wanted to show the image to demonstrate the lensing effect. Williams duly provided the image, and the result blew everyone's minds (plate 38). The image depicted *dozens* of finely distorted arcs in a dramatic tangential swirling pattern around the cluster, none of which could be seen with ground-based telescopes. Moreover, there were numerous multiple images—like the two images of the same quasar found in 1979.

Thanks to Hubble's exquisite image quality, these image pairs could easily be identified by virtue of their identical colours and shapes. Even without knowing the basics of gravitational lensing, people could immediately see the effect.

Kneib returned to Toulouse, and later Marseille, when I moved to California, but, through a visiting position, he came to Caltech in 2003. With him came his French student Johan Richard, who later became a postdoc at Caltech (chapter 8). Working with Kneib and Konrad Kuijken, a mathematically capable Belgian astrophysicist, we further developed the idea of looking for highly magnified distant galaxies using the Keck. The mathematics of gravitational lensing is elegant but non-trivial. Just as in conventional optics, when you look at something through a *gravitational* lens (e.g., a cluster of galaxies), the distorted image and its enlargement or magnification depends on the degree of alignment and relative distances to the intermediate lens (the cluster of galaxies) and the fainter, more distant background galaxy targeted for observation. In a remarkable, related effect, powerful gravitational lenses also define small peripheral zones called "critical curves" where the magnification can increase to huge factors. The location of these critical curves on the sky can be mathematically predicted for well-studied clusters depending on the redshift of a background source. Such sources seen along these critical curves can be explored as if one is using a telescope 20- to 100-times more powerful. So astronomers who know where to look have the ability to find hugely magnified distant sources at predictable redshifts. I continue to find this phenomenon inspirational: it's as if nature has provided us with its own giant telescope for free! Once in California, I set out to search the critical curves of many foreground lensing clusters for the most distant galaxies in the universe, effectively boosting yet further the power of the largest telescopes in the world.

In 2001, while using the Keck to scan the critical curves of our favourite foreground cluster lens Abell 2218, Kneib, Kuijken, and I found a distant galaxy at an unprecedentedly high redshift magnified 30 times. The observations to locate it and pin down its redshift were undertaken in two stages. During the first observing run we scanned the critical curve in a blind fashion, not knowing what we would find. We located a single emission line arising from hydrogen at a redshift $z = 5.576$. From its position Kneib calculated that there would be a second image of the same source on the other side of the critical curve. His geometric model was so accurate he could predict exactly where the counter-image would be now the redshift was known. On the second run, Kneib and I aligned the entrance slit of the spectrograph across both putative images. When the data came in from the computer, we immediately saw not one but two distinct emission lines at precisely the same redshift (plate 39). These "Eureka moments" are incredibly exhilarating, and we both rose from our chairs and did a "high five"!

Without the lensing magnification, this source was intrinsically so feeble it would have escaped detection in even the deepest Hubble Space Telescope images. One of the most distant sources discovered at that time, its actual unmagnified size is a mere 500 light years across, over 100 times smaller than the Milky Way. The discovery of this feeble tiny object, a mere million times the mass of the sun, was proof that our gravitational-lensing technique could find sources at the far reaches of the universe that we simply couldn't locate any other way. This success prompted us to propose surveying the critical curves of many such clusters, so that we could undertake a census of galaxies fainter than had ever been seen and in existence at a time when the universe was only 1 billion years old.

This critical curve survey became part of the thesis project of one of my first Caltech graduate students, Mike Santos, an energetic young man whose family came originally from Portugal. Together with Kneib and Kuijken, Santos and I used Keck to scan the critical curves in nine clusters, and we found a total of 11 feeble objects. Although none was more distant than that first found in Abell 2218, Santos's work gave us the first indication that for each luminous galaxy at high redshift, there was a much larger population of feeble ones.

The importance of Santos's thesis lay in the new light it shed on the demographics of galaxies in the early universe. As early as the 1960s, astronomers had noticed that the hydrogen gas in between galaxies is no longer in atomic form: the electrons have been stripped from their nuclear protons, and the gas has been "reionised." The origin of this "cosmic reionisation" was thought to be energetic radiation from early galaxies, but contributions from the then-known population of early galaxies seemed insufficient to produce this effect. At the time, theoretical astrophysicists speculated the existence of a much larger population of as-yet-undetected feeble galaxies that might do the job. Santos's observations provided a valuable and detailed glimpse of such a population. This topic is discussed further in chapter 8.

At this time a powerful new imaging camera, the Advanced Camera for Surveys, was installed on Hubble, and the staff at the Space Telescope Science Institute in Baltimore that operates Hubble decided that the cluster Abell 2218 would be a good target to demonstrate the instrument's remarkable new capabilities. From our earlier work, Kneib and I already knew the location of the critical curve for Abell 2218 as a function of redshift, so when the new image was released to the public, we immediately examined it and found a multiply imaged trio for which Kneib's model of the lensing geometry indicated the presence of a single source at a redshift $z \sim 7$. If correct, this would be the most distant known source, so we set out in the spring of 2003 to verify this redshift spectroscopically with Keck.

Mike Santos was in full flight on this quest, and what follows is a good illustration of how hard experimental research can be. First we used an optical spectrograph that would be sensitive to hydrogen emission up to a redshift of $z = 6.8$, but after an exposure of 6 hours we detected no signal. We returned a month later and used an infrared spectrograph that was sensitive to a hydrogen signal at redshifts beyond $z = 6.8$. After a marathon 9-hour exposure we detected a faint trace with a feature that we thought might indicate a redshift in the range $z \sim 6.8$ to $z \sim 7.0$, but we were not convinced everyone would believe it.

This is the dilemma of frontier science. The plain, sad fact was that we had no more observing time to explore our provocative, but tentative, new finding. Abell 2218 was now disappearing from view owing to the motion of the Earth around the sun and wouldn't be visible for another year. Santos was completing his thesis and would soon be on the job market. Should we write up this marginal result or defer publication until we could get more data? We could always apply for more observing time, but doing that for a "partially completed project" is a much tougher proposition than submitting a clever new idea. The Caltech TAC was bound to ask: "Will the additional time really improve upon the marginal result? And what if it doesn't?"

In such circumstances my philosophy has always been to be honest and proceed. We were exploring the edge of the known universe and had tried our best. We had a tentative result, and if correct it would be very exciting. Ideally we'd have had better data, but this is where we were. Accordingly, we submitted the paper, and, fortunately for us, it got a lot of attention—even leading to extensive media coverage at an American Association for the Advancement of Science meeting in Seattle. Soon after, we were contacted by Eiichi Egami, a Japanese colleague at the University of Arizona, who had followed up our target with the infrared Spitzer Space Telescope. His new data further supported our redshift and we enthusiastically worked with him and colleagues on a second article. Our paper indicated that the galaxy, in addition to being the most distant known object at the time, was relatively well established only 800 million years after the Big Bang. We argued that the galaxy had already been producing ionising radiation for 50 to 100 million years even at that early time. If representative, such a source and many others like it must have contributed significantly to cosmic reionisation.

While using my access to Keck to study gravitationally lensed sources, I was also undertaking a project with Andy Bunker, then at Exeter University in the United Kingdom, to track further examples of luminous objects based on Hubble images in areas of sky free from lensing—sometimes called "blank fields." This effort was successful in locating 10 sources at redshifts close to $z \sim 6$ that shone so brilliantly we could study them *without* the boost of gravitational lensing. High-quality images from Hubble were always crucial to our work. For the project with Bunker, the superlative quality of Hubble images taken in various filters was key to locating luminous high-redshift candidates using the "dropout" method introduced in chapter 6. For less luminous candidates viewed through lensing clusters, we also needed Hubble images so that Kneib could make models of the lensing geometry and match multiple images of the same background source.

The period between 2000 and 2004 was, without question, a golden era for Keck in this field, as the other 8-metre telescopes were only then beginning to operate. But, although I had the good luck to be in the right place at the right time, the competition was coming. The ESO's four 8.2-metre telescopes, named the Very Large Telescope (VLT), at Cerro Paranal in the Atacama Desert in northern Chile came online between 1998 and 2000, and Japan's Subaru 8.2-metre telescope atop Maunakea was operational by 1999. Inevitably, astronomers with access to these powerful new facilities began programmes similar to those being undertaken by me and my colleagues.

A particularly exciting apparent breakthrough occurred in 2004 when Roser Pelló, a Spanish astronomer working in Toulouse, France, announced the discovery of a redshift $z \sim 10$ galaxy from spectra taken at the VLT. A redshift of $z \sim 10$ corresponds to a time barely 450 million years after the Big Bang, so this finding, compared with my own result from the redshift $z \sim 7$ source in Abell 2218, represented a huge leap forward. Moreover, Pelló and her colleagues had used precisely our technique. She located the source along the critical curve of a cluster known as Abell 1835 and estimated the magnification to be a whopping 25- to 100-times, both in size and luminosity. The article summarising her team's finding concluded by emphasising how lensing alone would enable us to explore back to the epoch when galaxies emerged from darkness—that is, *cosmic dawn*.

An important aspect of any scientific endeavour is the reproducibility of key results. Pelló's data on her redshift $z \sim 10$ source were soon made available on a public archive operated by ESO, where they could be examined independently by others. Soon after, several astronomers posted critical articles on the internet indicating they could not reproduce her spectroscopic detection, and within a year or so the community was highly sceptical. Pelló was not alone in her ambiguous findings: my team later also presented some marginal candidates, including some at redshifts up to $z \sim 10$. While these were not refuted by others, our own attempts to verify them independently were unsuccessful. Pelló and I are hardly unique in this respect either. The pace of discovery is competitive, and, regrettably, this does sometimes lead to premature claims.

A more important discovery—and one that *was* widely acclaimed—came in 2006 from Japan. The Subaru Telescope is unique amongst 8- to 10-metre telescopes in having a wide field of view. This means it can take a picture of a field 16 times larger in area on the sky than either Keck or the VLT can. In surveying the sky for certain types of faint sources, it is therefore unparalleled. The Japanese decided to exploit this remarkable capability with purposely designed colour filters that can isolate distant galaxies emitting hydrogen emission lines at specific redshifts. The telescope's imaging area is so wide that, with this method, astronomers can find hundreds of high-redshift candidates in one exposure without any need for exhaustive spectroscopy.

Masanori Iye is one of Japan's most distinguished astronomers. Also a talented musician who distributes his own compositions on CD to fellow astronomers (including one entitled "Wishing on a Star"!), he played a key role in the construction of Subaru and pioneered its early use. In 2006 he located an emission line from a galaxy dubbed IOK-1 at a redshift of $z = 6.96$ (plate 40). Although his article was published a couple of years after the one in which we described our own discovery of a lensed source redshift $z \sim 6.8$ to $z \sim 7$ in Abell 2218, his Subaru imaging and subsequent spectroscopy clinched him the enviable target of the most distant known source.

Without question, my first 6 years at Caltech had been exhilarating. Several groups were now in the chase for the earliest galaxies, but I had two advantages: my access to the Keck telescopes and a team of really bright students and postdoctoral researchers. The benefits of searching through foreground clusters acting as gravitational lenses had been ably demonstrated. Looking ahead, the next challenge was to push even further into the period that we had begun to call the "reionisation era."

8

Entering the Reionisation Era

As a teenager in high school I remember being asked by a teacher to describe the two competing models of the universe to my classmates. One, familiar to most people today, is that the cosmos had a beginning— the Big Bang—which led to the expansion and evolution of the universe. However, a group of UK astronomers led by Fred Hoyle, a creative and scientifically unorthodox Yorkshireman, posited that the universe was, on average, unchanging. Unlike Einstein's struggles to maintain a static universe, prior to Lemaître's prediction of an expanding universe later endorsed by Hubble's data (chapters 2 and 3), Hoyle and colleagues didn't dispute the expansion of space. Instead they imagined a continuous creation of matter so that, as the universe expanded, the density of matter stayed the same. This model was dubbed the steady state theory. Unlike the Big Bang model, and as its name suggests, this posits that the universe on large scales never changes.

Fred Hoyle established what is now the Institute of Astronomy in Cambridge in 1967 when he was Plumian Professor, a position I later held. Although one of the most distinguished theoretical astronomers of his generation, who most famously demonstrated the origin of the chemical elements through various nuclear reactions in stars, he was often controversial. For example, in his later years he advocated that influenza viruses and even life itself originated in outer space and arrived on the Earth via comets and other alien bodies. Neither his admirers nor his detractors were terribly surprised when, just 5 years after creating his institute, he resigned over disagreements with the university. This led to many years in the academic wilderness and retirement in the Lake District, a region in northwest England famous for its rugged peaks and historic literary associations. Detached from the UK scientific establishment in which he was once well regarded, he focused on writing popular books, including science fiction, and giving lectures around the world.[1]

As the director of his former institute in the mid-1990s, I was keen to bring Hoyle back into the limelight, given how much he had done for UK astronomy. I was particularly grateful for his vision for the AAT, which had played a huge role in shaping my own career (chapter 4). So my Cambridge colleagues and I organised an event in honour of his 80th birthday in 1995 (plate 41), which was a splendid occasion. Even in the 1990s, against overwhelming evidence to the contrary, he still advocated the steady state theory—indeed, the very term "Big Bang" was invented by Fred himself as a derogatory expression in 1949! Accordingly, it was with some apprehension that I wrote to him ahead of the event to ask whether he'd like to give a lecture on the occasion of his birthday celebration. Of course, I feared he would present his non-standard view of cosmology, leaving me struggling to congratulate him in front of an embarrassed audience. I was somewhat relieved when, instead, he offered to present a talk entitled "The Shape of Tree Leaves," which turned out to be an entertaining display of fractal simulations he'd been doing on his elementary home computer. After the talk, when I asked if there were any questions, the distinguished crowd of Cambridge professors, postdocs, and students remained resolutely silent, until quantum physicist Lady Bertha Swirles Jeffreys (plate 41), aged 92 and sitting in the front row, asked brightly, "Can you do a horse chestnut?"

A major nail in the coffin of Hoyle's allegiance to the steady state theory was the serendipitous discovery made in 1964 by two young physicists, Arno Penzias and Robert Wilson. Working at the renowned Bell Telephone Company Research Laboratories in Murray Hill, New

---

1. There are several books describing Sir Fred Hoyle's life. A personal favourite on account of its scholarship and recognition of his success as a populariser of science is *Fred Hoyle's Universe* by Jane Gregory (Oxford University Press 2005). Fred's own personality can be gleaned in his (somewhat incomplete) biography *Home Is Where the Wind Blows* (University Science Books 1994).

Jersey, with a new radio antenna designed for communication purposes, the two scientists were perplexed by an incessant hiss emanating at microwave frequencies from all directions. Its origin remained a mystery until it was brought to the attention of a group of cosmologists working at nearby Princeton University, who had just submitted an article for publication predicting that the red-shifted thermal glow from the Big Bang should be a uniform microwave signal across the sky at a temperature of no more than 10 degrees above absolute zero. Their joint paper in *Astrophysical Journal Letters* detailing the discovery soon reached the general public in the form of a now-iconic May 1965 *New York Times* article that began, "Scientists at the Bell Telephone Laboratories . . . have observed what a group at Princeton University believe may be remnants of an explosion that gave birth to the universe." Subsequent space missions measured both the temperature of this *cosmic microwave background* (2.7 degrees above absolute zero) and the clustering pattern of intricate fluctuations in this temperature across the sky. Together with the well-established expansion of the universe, there seemed no escaping the fact that we live in an evolving universe.

Our modern picture of cosmic history was developed following Penzias and Wilson's remarkable discovery, which was awarded the 1978 Nobel Prize in Physics. In its early stages, the universe was a dense, hot plasma—a gaseous form in which hydrogen atoms were stripped of their negatively charged orbiting electrons. During this time, light particles (photons) were unable to travel very far without being scattered by these free electrons. However, as the universe expanded, the plasma cooled, and eventually the hydrogen nuclei (positively charged protons) and electrons slowed down sufficiently for their attractive electric forces to bind them together to make the hydrogen atom, a process called "recombination." Without free electrons, light rays were then able to travel freely in space, and this final scattering surface represents the origin of the microwave background. Calculations based on its temperature today and that above which hydrogen atoms are unable to retain their electrons indicate that the cosmic microwave background radiation emerged only 380,000 years after the Big Bang. This is a minute fraction of the current age of the universe, which is 13.8 billion years.

When Penzias and Wilson first detected this radiation, they were probing back 99.997% of the way to the beginning of time.

After the emergence of this primordial glow, the universe entered a period evocatively called the "dark ages," where these newly formed hydrogen atoms clumped under gravity into clouds, but there weren't any stars or galaxies.[2] As these clouds accreted more hydrogen gas, they became unstable and began to collapse. Just as an inflating bicycle tyre becomes hotter as the density of air increases within it, so these collapsing clouds eventually became sufficiently hot to ignite nuclear fusion, the process that makes the sun shine. The universe was bathed in starlight for the first time, a fundamental moment in cosmic history termed "cosmic dawn."

In 1965, soon after Penzias and Wilson's discovery, two Caltech graduate students, Jim Gunn and Bruce Peterson (introduced in my Australian adventures in chapter 4), realised that recently discovered distant quasars (chapter 3) could act as beacons that would reveal the nature of intergalactic gas—the gas between galaxies, consisting primarily of hydrogen. Just as a car's headlights can indicate the presence of fog ahead, so the spectrum of a distant quasar can trace the presence of hydrogen clouds in the direction towards the observer. Cold clouds of hydrogen distributed in intergalactic space would produce a significant atomic absorption signature in the spectrum of a distant quasar. Gunn and Peterson demonstrated that the absence of such an absorption signature in the most distant quasars known at the time (corresponding to a redshift $z \sim 2$) must indicate that the hydrogen in deep space is no longer in atomic form. They concluded that the hydrogen in the intergalactic medium is fully ionised and must, therefore, by some process have been returned to free protons and electrons, as was the case when the universe was very young. Gunn and Peterson postulated that this reionisation of intergalactic hydrogen, a process called "cosmic reionisation," must have occurred at a redshift

2. Astronomers are very fond of using the adjective "dark"; familiar examples include dark matter and dark energy. This adds an air of mystery and importance to the subject and can be used very effectively in securing research funding. The term "dark ages" was first coined by Martin Rees, the UK Astronomer Royal.

much larger than $z \sim 2$. For many years, theorists speculated that cosmic reionisation was initiated at cosmic dawn; that is, radiation from stars within the first galaxies was sufficiently energetic to begin breaking the electric bond that holds the hydrogen atom together (plate 42). Only in 2001, more than three decades after Gunn and Peterson's visionary article, did astronomers obtain the first observational evidence of when this transition from atomic to ionised hydrogen in intergalactic space occurred.

If, as theorists claimed, reionisation began at cosmic dawn, any information on its timing and duration would help observational astronomers like myself target our searches for the first galaxies. Following the discussion in Gunn and Peterson's early article, studies of the most distant quasars in the early 2000s revealed that by a redshift of $z \sim 6$ (1 billion years after the Big Bang), there was increased absorption along the line of sight by atomic hydrogen in the intergalactic medium. As such observations probed to earlier cosmic times, it seemed that more and more of the hydrogen in intergalactic space was in atomic form. Such observations therefore suggested that cosmic reionisation *ended* at around a redshift $z \sim 6$. But how much further back in time would we have to probe to find its *beginning*, perhaps initiated by the very first galaxies at cosmic dawn? And would our telescopes have the sensitivity to probe to such early times?

In 2003 the first results began to emerge from the latest microwave background satellite, the NASA-funded Wilkinson Microwave Anisotropy Probe (WMAP) named after David Wilkinson, a Princeton professor who sadly died just after the launch of the mission he had championed. WMAP presented the first measurements of the *polarisation* of the microwave background. Radiation can be polarised if it is scattered by electrons, in much the same way that sunlight can be polarised by aerosols in the atmosphere or reflections off a wet road. In the case of the microwave background, the polarisation is caused by the foreground electrons in intergalactic space (plate 43, *bottom*). Since these free electrons are present from the time reionisation began to the present day, analysing the polarisation signal and also the structure of the temperature fluctuations in their all-sky map, the WMAP team were able to provide the first approximate constraints on when reionisation occurred. Initially they found a surprising result: the scattering signal was so large that it implied reionisation began as early as a redshift $z \sim 20$.

I recall attending a lunch at the time with my astronomer colleagues in Caltech's faculty club, the Athenaeum, where those of us chasing high-redshift galaxies were despondent. We had just, with enormous difficulty, probed to a redshift $z \sim 7$ (chapter 7) and fresh off that achievement had begun to believe that with some additional effort we might soon reach far enough back in time to see cosmic dawn. Meanwhile, the WMAP result was telling us that this Holy Grail in our subject might lie as distant as a redshift $z \sim 20$ (a further 800 million years back in time). We'd have to wait maybe a decade or more before we had telescopes sufficiently powerful to peer that far back in time. As it happened, the first WMAP analysis overlooked a foreground contribution to the polarisation signal arising from spinning dust grains and cosmic ray electrons orbiting in the magnetic field of the Milky Way. In 2006, after correcting for this unrelated signal in an improved analysis, the WMAP team revised its estimate of when reionisation began to a period some 200 million years later. The European Space Agency's Planck mission, launched in 2009, further refined these measurements (plate 43, *top*), indicating that reionisation began at redshift $z \sim 12$, 400 million years after the Big Bang, and ended with a fully ionised intergalactic medium 600 million years later, by redshift $z \sim 6$. This was more encouraging and was possibly within reach of our present telescopes with renewed effort. The developing picture of the early universe, courtesy of WMAP and Planck, gave a huge impetus to distant-galaxy explorers like myself.

Hydrogen is the most abundant element in the universe. Along with a smaller amount of helium, it is most of the material from which stars are formed and comprises the nuclear fuel that makes them shine. When hydrogen gas is heated by young stars, it glows in a prominent spectrum line of hydrogen called the Lyman-alpha line. A "spectrum line" represents radiation emitted at a fixed wavelength according to the atomic energy levels occupied by electrons in the quantum theory of physics. The observed wavelength of such a recognisable line in a distant galaxy gives us its redshift and hence a time stamp of when that galaxy emitted its radiation. To recap the closing events of the last chapter, in 2006 the Japanese astronomer Masanori Iye had discovered his famous galaxy IOK-1 at a redshift $z = 6.96$, by affixing a special filter expressly designed to seek out this Lyman-alpha emission line on the Subaru Telescope. The critical curve survey I undertook with Mike Santos and "blank field" campaign with Andy Bunker also yielded redshifts based on detecting Lyman-alpha emission.

However, beyond a redshift $z \sim 7.1$, Lyman-alpha emission leaves the wavelength range of optical spectrographs and moves into the near-infrared, where the necessary detectors are more expensive and, in the late 2000s, were not at the state-of-the-art level of their optical counter-parts. Modern optical spectrographs use charge-coupled devices (CCDs, introduced in chapter 2), which are highly efficient and have been available in large formats since the late 1980s. However, the silicon materials that form the basis of optical CCDs cannot function at longer wavelengths. The material of choice for near-infrared detectors is usu-ally a compound of mercury, cadmium, and telluride (HgCdTe) or one of indium and antimonide (InSb). Only a few companies worldwide have developed the technologies to manufacture these detectors, and, in the late 2000s, a large-format infrared detector could set an astron-omy programme back $250,000.

In addition to the fact that near-infrared-detector technology lagged behind that of the optical, the glow from the Earth's atmosphere is an enduring nuisance in infrared astronomy. Even at night, the atmosphere radiates by various processes reflecting both its temperature and various reactions between it and incoming cosmic particles. At the low atmo-spheric temperatures familiar to anyone who has seen frost on the win-dow of a long-distance jet aircraft, the thermal glow becomes significant only in the near- and mid-infrared spectral regions. There are also nu-merous sky emission lines at particular wavelengths arising from the oxygen-hydrogen radical, a molecule with one unpaired electron cre-ated by interactions between hydrogen and ozone in a layer about 90 kilometres high. These annoying OH lines, which occur only in the red and infrared spectral regions, pepper the spectrum of a faint galaxy and cannot be perfectly removed in software. The only way to fully avoid these atmospheric issues that plague infrared measurement is to go into space.

A succession of two enterprising Caltech PhD students worked with me as I contemplated probing the reionisation era beyond a redshift $z \sim 7$. Dan Stark came from Wisconsin in 2004, and Matt Schenker from New York City arrived in 2009. Keck's near-infrared spectrograph during Stark's tenure was called NIRSPEC (Near Infrared Spectrograph). Built at the University of California at Los Angeles by Ian McLean, a self-effacing but very talented Scot whom I knew well from our UKIRT days and a pioneer in near-infrared instruments, NIRSPEC had a 2048 by 2048 pixel InSb detector. Using NIRSPEC, Stark and I began a long and arduous effort to search for Lyman-alpha emission in the reionisation era. From the start, I realised that this was a somewhat risky undertaking. While it was true that Iye's galaxy at $z = 6.96$ had prominent Lyman-alpha emission, the technique he had used to discover it relied on searching first with a filter expressly designed to detect the line. He would have entirely missed any galaxies that did not emit the line. How sure could Stark and I be that all early galaxies would obligingly exhibit this line? Recall also that the lensed redshift $z \sim 6.8$–7.0 galaxy in Abell 2218 I studied with Mike Santos and Jean-Paul Kneib had shown no evidence of Lyman-alpha emission (chapter 7).

To say our initial attempts to probe beyond redshift $z \sim 7$ were challenging is an understatement—we had enormous difficulties. We began with what, in retrospect, was an overly ambitious attempt to use NIR-SPEC to scan the critical curves of nine clusters, searching for Lyman-alpha emission in the redshift range $z \sim 8.5$ to $z \sim 10.4$. Over the course of four Keck observing runs from 2004 to 2005, we tentatively put forward six possible candidates. None was particularly convincing given the difficulties of subtracting the foreground sky signal, and during the observing we were never quite sure whether we were maintaining an accurate position on our targets for the necessarily long exposures of 3–4 hours. As we hesitantly admitted in a lengthy paper published in *Astrophysical Journal* in 2007, "one might conclude we have reached new territory where we may never quite know with certainty whether an object is at (redshift) $z \sim 10$."[3] Even worse, our attempts to confirm some of our NIRSPEC candidates using an independent infrared spectrograph on the Subaru Telescope were not convincing.

During one of these observing runs, the journalist Michael Lemonick from *Time* magazine came to Hawaii to watch Stark and me observe. Inviting him along seemed a good idea when we began our ambitious survey, but his article "How the Stars Were Born" (September 4, 2006), which was even featured on the cover of the magazine,[4] does not paint a very flattering picture of our observing skills. It opens with a description of me pacing up and down in the Keck control room, cursing that the telescope won't focus and that the night is lost. It turns out that, unknowingly in the afternoon, we had entered a spurious character on the keyboard when submitting information into the Keck computer; somehow this sent the telescope into a trance. It took the Keck engineers 2 hours to figure out what was going on!

Undeterred, we began a second series of observing runs—led by postdoc Johan Richard, a soft-spoken Frenchman introduced to us by Kneib—which was equally challenging. Here, we analysed deep Hubble images of six clusters, using photometric redshift estimates (i.e., based on colours, see chapters 5 and 6) to identify faint lensed galaxy candidates beyond redshift $z \sim 7$. We located 12 examples and tried to secure spectroscopic redshifts based on Lyman-alpha emission for 7 of these candidates with Keck in 2007. No lines were detected.

It was hard to maintain enthusiasm within my group at this time. We had devoted a significant fraction of my Keck time to this project for 3 years, and had not convincingly confirmed with a spectrum a single candidate beyond redshift $z \sim 7$. We were pushing NIRSPEC to new limits and continued to worry whether our targets were drifting out of its entrance aperture. It seemed an impossible task. Everything had to

3. A Keck Survey for Gravitationally Lensed Lyman Alpha Emitters in the Redshift Range $8.5 < z < 10.4$: New Constraints on the Contribution of Low-Luminosity Sources to Cosmic Reionization, D. P. Stark, R. S. Ellis, J. Richard, J.-P. Kneib, G. P. Smith, and M. R. Santos, *Astrophysical Journal*, vol 663 (2007), pp10–28.

4. M. Lemonick, "How the Stars Were Born," *Time Magazine*, September 4, 2006, 42–51. See the front cover, https://content.time.com/time/covers/0,16641,20060904,00.html.

be perfect: good weather on the night, accurate telescope guiding to ensure we remained on target for several hours, reliable sky subtraction, and careful statistical analyses. Over the years, given such challenges, I had learned the dangers of having graduate students rely too heavily on a single project for their thesis. In Stark's case, he and I had prudently designed a parallel programme aimed at determining approximately how many Lyman-alpha emitting galaxies could be detected in the redshift range $z \sim 3$ to $z \sim 7$. This less daunting project could be accomplished using a wide-field optical Keck spectrograph called DEIMOS (the Deep Imaging Multi-Object Spectrograph). Built at the University of California, Santa Cruz, by a team led by the noted astronomer Sandra "Sandy" Faber (introduced briefly in chapter 6), DEIMOS had the potential to secure the information we sought, admittedly at lower redshifts, for hundreds of faint galaxies.

Faber has a towering reputation in many aspects of astronomy (pioneering observations, theoretical implications, and revolutionary instrumentation). I must admit I have often found her a bit daunting. I recall that the first time we met she looked at me with piercing eyes and asked if I could describe, succinctly, what fundamental result I had achieved in my PhD thesis. The honest answer was not a lot! When I arrived at Caltech in late 1999, she was nonetheless keen to welcome me to California. We worked together to consider whether DEIMOS could be upgraded to include an infrared capability, but this turned out to be impractical. She also proposed that Caltech and the University of California, Santa Cruz, should join forces to exploit her DEIMOS instrument and undertake a very large redshift survey of galaxies. Her idea was that Chuck Steidel and I would apply for half the required observing time through the Caltech TAC and she would apply for the remainder via the University of California TAC. Although I welcomed this idea, I failed to realise that my colleagues fiercely guarded the sovereignty of Caltech's observing time on the Keck telescopes and saw this Caltech–University of California partnership as a way for Faber to gain access to Caltech's share of Keck time. As a result, the partnership was abandoned and Faber went ahead with her own survey, called DEEP2, which over several years secured spectra for almost 53,000 galaxies to and beyond a redshift $z \sim 1$. DEEP2 was the ambitious next step in exploring the spatial distribution and evolutionary properties of galaxies following the more local two-degree field survey my colleagues and I conducted at the AAT (chapter 4). The DEEP2 survey represented a milestone in evolutionary studies of galaxies over the past 7 billion years of cosmic history and led to over 150 scientific articles.

Returning to Dan Stark's campaign, targeting around 600 galaxies in the redshift range $3 < z < 7$ selected from Hubble images using the same colour selection method exploited by Chuck Steidel (see chapter 6 and plate 34), Stark used DEIMOS to determine how many revealed Lyman-alpha emission. On one observing run, we found Lyman-alpha emission in more than 70 galaxies at redshift $z \sim 6$, a dramatic improvement over the earlier work I did with Andy Bunker (chapter 7). Dan's extensive redshift survey complemented more-detailed spectroscopic studies of over 800 similarly selected galaxies at redshifts $z \sim 2–3$ by two of Steidel's graduate students, Alice Shapley and Dawn Erb. The collective data set, remarkably assembled entirely by Caltech students, provided the first comprehensive view of spectral evolution in star-forming galaxies across a broad redshift range spanning 2 billion years of early cosmic history.

Out of this success story arose a new puzzle. Stark's impressive achievement in recovering many examples of Lyman-alpha at redshift $z \sim 6$ indicated that we should have readily seen the line in our more distant surveys beyond redshift $z \sim 7$. He suggested there might be a physical reason for our lack of success. During the reionisation era, any intergalactic atomic hydrogen clouds would act as a "fog," scattering and diffusing the Lyman-alpha emission produced in galaxies. This dimming of Lyman-alpha emission by atomic hydrogen in intergalactic space was first pointed out in an article by Jordi Miralda-Escudé and Martin Rees in 1998. Stark argued we might be able use the *fraction of galaxies at a given redshift that revealed Lyman-alpha* to determine how much hydrogen in the intergalactic medium at that time was in either atomic or ionised form. As we entered the "foggy" reionisation era, the fraction of Lyman-alpha-emitting galaxies should decline, owing to the dimming predicted by Miralda-Escudé and Rees.

FIGURE 8.1. Dan Stark (left) and Matt Schenker at the Keck Observatory during the challenging NIRSPEC era.

It was truly inspirational working with Stark (figure 8.1). Despite two extremely arduous and seemingly unsuccessful campaigns with NIRSPEC, he maintained his dedication, triumphed with his DEIMOS observations, and recognised that our failure to detect Lyman-alpha beyond redshift $z \sim 7$ might be due to the evolving state of the intergalactic medium. He completed his dissertation with a flourish and moved to a postdoctoral position in Cambridge.

Fortunately for me, Stark was succeeded by Matt Schenker (figure 8.1), who arrived at Caltech with an undergraduate degree from Dartmouth College. Inspired by the idea of using the visibility or otherwise of Lyman-alpha emission to gauge the ionisation state of the intergalactic medium, Schenker and I used NIRSPEC to target 19 colour-selected galaxies thought to lie in the redshift range $z \sim 6.3$ to $z \sim 8.8$, carefully matching the luminosities of these targets to the larger, lower-redshift DEIMOS sample constructed by Stark. This time we undertook a much more systematic approach, exposing on each galaxy so we could confirm or rule out Lyman-alpha emission to a predetermined level. In contrast to Stark's 50% success rate in detecting this line at redshift $z \sim 6$, out of the 19 higher-redshift galaxies, only 3 showed Lyman-alpha, a success rate of less than 16%. We concluded the visibility of the line must decline

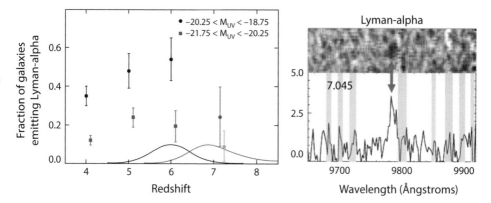

FIGURE 8.2. Charting the end of reionisation. (*Left panel*) The rise and fall in the fraction of galaxies showing the Lyman-alpha emission line of hydrogen with increasing redshift, supporting Dan Stark's suggestion that the line is suppressed as we enter the reionisation era beyond redshift 6. (*Right panel*) The Lyman-alpha line from the lensed galaxy A1703-zD6 at $z = 7.045$, breaking the redshift $z = 7$ barrier.

rapidly beyond a redshift $z \sim 6$, most likely because the intergalactic medium was not yet fully ionised and, therefore, that we were entering the era when reionisation was still underway. A satisfying by-product was confirmation of a galaxy at a redshift $z = 7.045$ (figure 8.2). After 7 years of trying, we had finally broken the "redshift 7 barrier"!

As is common in the fast-paced world of extragalactic astronomy, Stark's article describing how the abundance of Lyman-alpha emission in carefully controlled samples of high-redshift galaxies could provide a valuable probe of when reionisation ended was posted on the internet in March 2010, ahead of its formal publication in *Astrophysical Journal* later in November. The paper included an early version of the trend posted later on the internet in October 2011 following additional data secured by Matt Schenker. By this time, however, we had competition. A group in Rome headed by Adriano Fontana and Laura Pentericci had embarked on a similar campaign, using the ESO VLT. It was gratifying to see that their first findings, posted in October 2010, confirmed the decline in Lyman-alpha visibility beyond $z \sim 6$ seen by Stark and Schenker. However, their optical spectrograph could probe the visibility of Lyman-alpha only to a redshift $z \sim 7.1$, as opposed to the wider baseline to redshift

$z \sim 8.8$ possible with NIRSPEC. Nevertheless, they raced ahead, and in February 2011 their team broke the redshift 7 barrier with two sources at redshifts $z = 7.008$ and $z = 7.109$, ahead of our own announcement in October 2011. Remarkably, the Italian record was broken only a few weeks later when a Japanese team led by Masami Ouchi and Yoshiako Ono posted the Keck spectrum of a galaxy at a redshift $z = 7.21$.

Clearly, cosmic reionisation was becoming a very hot topic! Three independent groups had found their first galaxies beyond a redshift $z \sim 7$ in a single year, and international astronomy conferences and workshops were abuzz discussing the implications. In July 2011, just after announcing our latest results, I returned to Cambridge for a meeting entitled "New Horizons for High Redshift" attended by all the key high-redshift players—observers and theorists. Two questions dominated the proceedings: When did reionisation begin and end, and were young galaxies primarily responsible? Assigned the task of giving the conference summary, I expressed optimism that the numerous, widespread studies of Lyman-alpha emission in galaxies and hydrogen absorption in quasar spectra were pointing to reionisation ending at about redshift $z \sim 6$. I then went on to emphasise that there were still many uncertainties about the role of young galaxies in driving the process. To address the second question, we needed to know the number of galaxies per unit volume at early times, their luminosity distribution, and the extent to which each galaxy at that epoch was capable of generating and releasing into the intergalactic medium sufficient amounts of ionising radiation. As usual, theorists were happy to speculate, but the cold hard fact was there was little reliable observational data on the demographics and intrinsic properties of galaxies in the reionisation era. Breaking a redshift record might be good for morale, but it wasn't answering the questions being discussed in the community.

Here it becomes necessary to step back a couple of years and introduce a parallel theme in exploring the demographics of galaxies in the reionisation era through long-exposure images taken with the Hubble Space Telescope. In May 2009, 2 years before my guardedly optimistic conference summary at Cambridge, mission specialists aboard the Space Shuttle *Atlantis* had installed a new and improved imaging camera on the HST. Known as the Wide Field Camera 3 (WFC3), it offered an enlarged field of view and better sensitivity than its predecessors, particularly in the near-infrared band. Following Bob Williams's initiative with the Hubble Deep Field imaging programme (chapter 6), his successor as director of the Space Telescope Science Institute, Steve Beckwith, continued this theme with an even deeper Hubble Ultra-Deep Field (HUDF) campaign in 2004, representing a total exposure time of over 250 hours with the optical Advanced Camera for Surveys. In 2009 this HUDF data set was augmented further with the first deep-infrared images taken using WFC3 by Garth Illingworth at University of California, Santa Cruz, who beat our own proposal for similar science and emerged as our primary competitor for the next few years.

Garth Illingworth was a formidable opponent. A jovial and burly Australian, he led a large team of self-styled "commandos" including Pieter van Dokkum, a brilliant Dutch professor at Yale, and two postdocs, Rychard Bouwens—a delightfully eccentric and talented individual often regarded as the joker at conferences—and Pascal Oesch, an unassuming Swiss researcher with an efficiency and reliability characteristic of the trains in his native country. In addition to his knack for attracting talent, Illingworth was remarkably well connected. Respected in the upper echelons of NASA, he frequently chaired national space science committees and could readily call on technical support from those who understood how to maximise the performance of Hubble's instruments. As discussed in chapter 7, Caltech astronomers were often mistrusted elsewhere in the US community because of their uniquely generous access to the Keck telescopes. Sometimes I sensed that Illingworth had many supportive colleagues at Santa Cruz who would not be at all sorry to see Richard Ellis of Caltech beaten out in the competition for precious Hubble observing time.

To indicate the intensity of the competition in exploring the reionisation era, when Illingworth's near-infrared HUDF data were publicly released in September 2009, within only 2 months nine separate papers from various groups were posted on the internet. Teaming up with colleagues at the Royal Observatory, Edinburgh, led by Jim Dunlop and Ross McLure, Dan Stark and I found 15 redshift $z \sim 7$ and 3 redshift $z \sim 8$

candidates based on their infrared colours in this remarkably deep image. Similar results were found independently by Illingworth's young quarterbacks Bouwens and Oesch. At this point it is important to place these advances in context. None of these newly revealed $z > 7$ galaxies had spectroscopically confirmed redshifts; their redshifts were *estimates*, derived simply from their colours. In some sense this was a return to counting distant galaxies, using Hubble's powerful new camera but with the added trick of using Steidel's "dropout" colour selection method (chapter 6) to approximately determine their redshifts.

Nonetheless, Hubble had potentially opened up the universe to redshift $z \sim 8$, well into the heart of the reionisation era according to the Planck results. Two key results emerged: Firstly, the abundance of starforming galaxies fell off rapidly with increasing redshift, by a factor of over 10 from redshift $z \sim 4$ to $z \sim 8$, within a time interval of 900 million years. Extrapolating from this trend suggested that by a redshift $z \sim 12$ galaxies were virtually non-existent, perhaps a harbinger of cosmic dawn. Secondly, it seems clear that Mike Santos had been correct when he conjectured, nearly a decade earlier, that feeble low-luminosity galaxies were largely responsible for generating ultraviolet light at early times. Illingworth's deep image suggested that at each approximately determined redshift, feeble galaxies were far more numerous than their luminous counterparts.

However, although four independent teams (including ours) may have agreed on the *number* of redshift $z \sim 7$–8 galaxies in Illingworth's data, as the results came under closer scrutiny, disagreements arose as to *which ones they were*. This was, to put it mildly, a source of considerable embarrassment within the high-redshift community; an outsider would wonder whom to believe. Jim Dunlop at Edinburgh convinced our team that Illingworth's observing strategy with Hubble was inadequate, and that to reliably push beyond $z \sim 8$ it was necessary to double the exposure time Illingworth had employed in one key infrared filter, and also bring into play a further, additional, infrared filter that would ensure we could confidently distinguish redshift $z \sim 7$ to 8 galaxies from those expected at redshifts $z \sim 8$ to 10. I had known Dunlop well since the 1980s. A Scotsman with the obligatory strong accent, he began his career as a radio astronomer but soon emerged as the most successful "multi-wavelength" UK astronomer of his generation. Dunlop combines scientific vision and leadership with a rare eye for technical detail. Thanks to his creative thinking, after two unsuccessful attempts we finally outdistanced Illingworth and his colleagues in the competition for HST time and secured 128 orbits of Hubble time to probe even deeper in the HUDF in 2012 (HUDF12; plate 44).

Writing a competitive Hubble proposal is an enormous investment of team effort, and our eventual success was the result of a partnership between my Caltech team (both existing and former members such as Stark and Schenker) and strong leadership from Dunlop and McLure and others in Edinburgh. Failed proposals are a fact of life in the cut and thrust of big science, but in 2010 we were able to turn our earlier setbacks to advantage by publishing the basis of our unsupported proposal in a widely read article "Early Star-Forming Galaxies and the Reionisation of the Universe" in the journal *Nature*.

Our lead author on that article was Brant Robertson, one of the smartest people I've ever worked with. Although a theoretical astrophysicist by training (with a PhD from Harvard), unusually for a scientist of his stripe he seemed to relish the idea of working with observers. Having worked with many theorists who questioned why I felt the need to go observing as they already knew the answers to most of my questions, I found his attitude uniquely refreshing. Time and again I admired Robertson's ability to undertake complex calculations across a wide range of astrophysical problems and, impressively, work things out himself, rather than taking his inspiration from someone else's results. He was also first author on a subsequent article that summarised the scientific outcome of our successful HUDF12 campaign, whose major achievement was to push further into the reionisation era to redshifts $z \sim 10$ for the first time. Specifically, based on their colours, we located seven candidates beyond a redshift $z \sim 8.5$ (plate 44, *bottom*).

A synthesis of what we learned from HUDF12 in the context of the history of reionisation as determined by the Planck microwave background data was published in a further article by Brant Robertson, myself, Jim Dunlop, and Steve Furlanetto (a US theorist in our team) in

2015. As discussed above, the Planck team determined that the amount of foreground electron scattering was much less than that indicated initially by the earlier WMAP satellite. The team used its measurement to infer the likely beginning and end of the reionisation era. Using HUDF12, we now had a meaningful census of the abundance of galaxies through much of their claimed reionisation era, and, making some assumptions about their ionising capability, we were able to reconcile our data with the Planck measurement (plate 45). Our article concluded that "our analysis strengthens the conclusion that star-forming galaxies dominated the reionisation process."[5]

This period of my career was very satisfying. Working with Matt Schenker and Brant Robertson at Caltech was very productive and the connection with the Edinburgh group was central to our successful HUDF proposal. I believed we had demonstrated that it was quite likely that star-forming galaxies were the principal drivers of cosmic reionisation, even though some of their properties—such as the strength of their ionising radiation—remained to be pinned down. Jim Dunlop kindly arranged for a Carnegie Centennial professorship held at Edinburgh, funded by the Carnegie Trust for the Universities of Scotland. This enabled me to spend a month each summer at the Royal Observatory. Andrew Carnegie, who did so much for astronomy in the early twentieth century, continued to leave his legacy on the subject 100 years later.

Our findings notwithstanding, the Illingworth empire was hardly ready to throw in the towel after losing the competition for HUDF12. Following our publications, Illingworth's group produced its own series of articles based upon our data and subsequent Hubble and Spitzer Space Telescope observations. However, an interesting turn of events followed our analysis of the only galaxy we found beyond a redshift $z \sim 8.5$, one that the Illingworth team had originally detected in its 2009 data. This enigmatic galaxy, unromantically designated UDFj-39546284, was initially claimed by Illingworth and colleagues to be at a redshift

5. Cosmic Reionization and Early Star-Forming Galaxies: A Joint Analysis of Constraints from Planck and the Hubble Space Telescope, B. E. Robertson, R. S. Ellis, S. R. Furlanetto, and J. S. Dunlop, *Astrophysical Journal Letters*, vol 802 (2015), L19.

$z \sim 10$. Although this result was published in *Nature* with the usual fanfare, many, including myself, were sceptical. It was visible only in a single infrared Hubble filter; there was no other information. It could easily be a galaxy at lower redshift with an intense emission line lying within the wavelength window of this filter. Rychard Bouwens and colleagues even had the courage to estimate the volume density of galaxies from this single candidate, claiming a value a factor of 10 lower than that seen at redshift $z \sim 8$. Looking back a further 170 million years before a redshift $z \sim 8$, they claimed the galaxy population had dramatically decreased. Their paper concluded "The 100–200 million years before (redshift) $z \sim 10$ is clearly a crucial phase in the assembly of the earliest galaxies."[6]

I confess I was among the unconvinced. It seemed to me that proclaiming a density of galaxies based on a single source was like walking into a cow pasture, happening upon an ancient Roman coin, and using that discovery to calculate the total number of such coins across the whole of England! As our deeper HUDF12 data gradually became public, they naturally confirmed the presence of UDFj-39546284. True to form, Bouwens and colleagues immediately posted a hurried confirmation article on the internet before we had even submitted our own results from HUDF12.

Recall, however, that our programme also involved observing in a new, additional, colour filter that Jim Dunlop had insisted was necessary to robustly confirm the presence of galaxies beyond $z \sim 8.5$. Unbeknown to Bouwens, the source was *not seen in the new filter*, which ruled out a galaxy at a redshift $z \sim 10$. Either UDFj-39546284 was a foreground object, as many of us suspected when it was first discovered, or it would have to lie at an even higher redshift of $z \sim 11.9$. A spectrum published by Gabriel Brammer, a member of Illingworth's team, in 2013 revealed a marginal emission line most likely due to oxygen at a redshift $z = 1.605$.

Notwithstanding their premature claim on the nature of UDFj-39546284, the Illingworth team continued to maintain that the number

---

6. A Candidate Redshift $z = 10$ Galaxy and Rapid Changes in That Population at an Age of 500 Myr, R. J. Bouwens, G. D. Illingworth, I. Labbé, P. A. Oesch, M. Trenti, C. M. Carollo, P. G. van Dokkum, M. Franx, M. Stiavelli, V. González, D. Magee, and L. Bradley, *Nature*, vol 469 (2011), p504.

of galaxies in the early universe declined sharply beyond redshift $z \sim 8.5$. Despite additional Hubble investigations, such as the Cosmic Assembly Near-infrared Deep Extragalactic Legacy Survey (CANDELS), led by Sandra Faber at Santa Cruz and Harry Ferguson at the Space Telescope Science Institute, which targeted a wide area of sky, and various campaigns to identify gravitationally lensed galaxies to $z \sim 11$, the question remained unresolved. From 2012 to 2015, Oesch and Bouwens utilised these surveys to provide improved measures of the declining abundance of galaxies. However, Dunlop and his Edinburgh student Derek McLeod continued to refute the suggestion of a rapid decline in the abundances beyond redshift $z \sim 8$. Although the numbers of early candidates continued to increase, the statistical uncertainties were still significant, and by 2015 there was still not a single convincing spectroscopic redshift beyond $z \sim 8$.

With mixed feelings, in August 2015 my wife and I left Pasadena. Our growing UK family was based in London and it was time to enjoy our grandchildren. I had decided to take a 2-year position at the ESO headquarters near Munich and then return to my alma mater, UCL, in 2017. After 16 years at Caltech, the landscape of high-redshift galaxies had changed beyond recognition. Collectively, our team; Garth Illingworth's; and those in Japan, Italy, and elsewhere had pushed the galaxy frontier back from a redshift $z \sim 4$ to potentially a redshift $z \sim 11$, spanning most of the epoch of reionisation according to the Planck team's interpretation of its data. Even as rivalries and arguments about the interpretation of the data continued, centring chiefly on the validity of claims for a rapid decline in galaxy numbers beyond redshift $z \sim 8$, we certainly seemed prepared to close in on cosmic dawn.

9

The Arrival of the Euro

A Glimpse of Cosmic Dawn

The European Southern Observatory (ESO) is astronomy—big time. There are many aspects of ESO that place it at the forefront of world astronomy. As a treaty organisation between 16 national governments, its annual budget is large and secure, so it can effectively plan new telescopes and instruments and manage the world's largest observatory. With its headquarters in Garching near Munich in Germany, it commands respect from Europe's large industrial companies, who partner in building new capabilities. It is frequently argued that "big science" in astronomy requires a business model equivalent to that in high-energy physics, where European nations join forces at the Centre for European Nuclear Research (CERN) to exploit the world's largest particle physics laboratory. Indeed, ESO has built four 8-metre telescopes, each costing more than 100 million euros, equipped with many state-of-the-art science instruments. This ambition and confidence has generated a culture whereby international teams of as many as 50 astronomers, often spread across the member states, conduct grand research projects that require upwards of 20 nights of observing time. The international teams that build the science instruments, each typically costing 20–30 million euros, are rewarded with "guaranteed time," which over a few years can amount to several hundred nights of observing time. This is a far cry from a lone professor building a spectrograph in a university laboratory and going for an observing run with a couple of students.

The "S" in ESO is because of Chile, where all its numerous telescopes are operated. A remarkable country, 4300 kilometres long and only 350 kilometres across at its widest point, Chile has a variety of climates. The cold current discovered by Alexander von Humboldt, together with the rapid rise of the Andes mountains from the coast north of Santiago, provides a barrier to humidity across the land that ensures an exceptionally dry climate. In some parts of the Atacama Desert it has never rained in recorded history.

As an organisation, ESO was first envisaged following a meeting between Walter Baade, a German astronomer who emigrated to the United States in 1931 (chapter 3), and Jan Oort (1900–1992), a distinguished Dutch astronomer at Leiden University. Oort was encouraged by Baade to rally European collaborative efforts in funding a southern large telescope, which led to an international partnership agreement in 1954 and the commencement of site testing in South Africa in 1955.[1] However, the political situation in South Africa, and news of spectacular sites being tested in Chile by the Association of Universities for Research in Astronomy (AURA), a collection of American universities, forced an uncomfortable rethink. The individuals who had meticulously tested mountain sites in South Africa for 8 years, camping out with meteorological equipment and small telescopes, must have been pretty disappointed when ESO—then comprising only five national member states—decided in 1963 to focus instead on Chile.

Unusually for a European, Oort was promoting the advantages of cooperation with the Americans. The suggestion was to share facilities on a site AURA had purchased that encompassed the peaks of Cerro Tololo, Morado, and Pachón near the coastal town of La Serena (plate 46). But it was not to be. ESO was established from the outset as an intergovernmental organisation similar to CERN, whereas AURA represented a partnership between academic institutions. ESO sought a

---

1. The United Kingdom was initially part of this cooperative endeavour but withdrew in 1961 to focus on its collaboration with Australia. It seems Richard Woolley, the Astronomer Royal (chapter 1), was particularly influential in this decision. One wonders how UK astronomy would have developed had the country joined ESO in 1963. As it happened, the United Kingdom flourished outside ESO for the next 30 years.

direct arrangement with the Chilean government and wanted to own its site, not lease one from AURA. In a swift but controversial move, Otto Heckmann (1901–1983), the first ESO director-general, personally negotiated the purchase from the Chilean government of a separate mountain called La Silla, 200 kilometres further north, and established an operational base in Santiago. Both AURA and Oort were disappointed at the loss of a collaborative opportunity. Even ESO's council members were initially upset by Heckmann's unilateral action. AURA had undertaken a comprehensive site-testing campaign from 1958 to 1962, meticulously recording the observing conditions on eight mountains ranging from the vicinity of Santiago up to Copiapo, 800 kilometres north into the Atacama; La Silla was not included in this list of sites, and so was an unknown prospect.[2]

I met Jan Oort briefly in the late 1980s when I was visiting the Space Telescope Science Institute in Baltimore. The Hubble Space Telescope was shortly to be launched, and the first observing programmes on this unique space telescope had already been selected. There was this distinguished octogenarian sitting in the library reading the chosen science cases. He died in 1992, shortly after the eventual (delayed) launch of Hubble.

In the 1980s when I was at Durham using the Anglo-Australian, William Herschel, and UK Infrared telescopes, ESO's facilities were entirely based at La Silla, a sprawling site containing a range of 0.5- to 4-metre telescopes. Although I visited fairly often, I think it fair to say British astronomers did not consider ESO presented much of a threat. There was perhaps even a smugness in some quarters of the UK (and Australian) astronomical communities. On one occasion I was invited to observe with the recently completed New Technology Telescope on

2. The man who led AURA's site-testing campaign (and became the first acting director at Cerro Tololo) was Jürgen Stock, an enigmatic German national who did more than anyone to promote the astronomical prospects of Chile. He moved to Venezuela in 1970. By a remarkable coincidence, I had a collaborator who worked in Venezuela, and so, in the late 1980s, I met this modest man whose contributions to Chilean astronomy were fundamental. As is so often the case when meeting an older figure, I didn't at that time recognise the significance of his earlier contributions. Stock died in 2004.

La Silla, largely to comment on whether things were improving at ESO. UK astronomers were seemingly held in high regard within the ESO organisation.

In this context it is interesting to read an account of ESO during its formative years by Lodewijk Woltjer, the director-general from 1975 to 1987 introduced in chapter 5. Woltjer is surprisingly candid, contrasting George Ellery Hale's accurate and ambitious scientific vision for Palomar in 1928 with the parochial view of European astronomers almost 50 years later. In Woltjer's words, "Few things happen very fast in Europe."[3] Woltjer deserves great credit for initiating a major change of direction, launching the Very Large Telescope (VLT) project in 1987, destined for a site of his choice, Cerro Paranal, 600 kilometres north of La Silla. The VLT comprises four 8.2-metre telescopes that can be used independently or together as an interferometer for exquisitely high-resolution studies of bright sources. These four giant telescopes were commissioned as planned between 1999 and 2001. From that point on, nobody ridiculed ESO, not even the Americans. Recognising this, the United Kingdom belatedly joined as an ESO member in 2002. Now, in addition to serving astronomers from 16 European countries, there are additional agreements with Chile and Australia. ESO serves almost half the professional astronomers in the world.

In 2018, my photographer Julian Abrams and I undertook a marathon drive of over 1200 kilometres from Santiago up the Panamericana Norte to Cerro Paranal. We drove for ages through barren landscapes, including a long stretch along the Pacific coast north of the port city of Caldera, which serves the vast mining activity in the Atacama Desert. Paranal is at an altitude of 2650 metres, and less than 15 kilometres as the crow flies from the Pacific Ocean. Although the coast was shrouded in fog, as we climbed inland we soon emerged into brilliant deep blue skies and sunshine.

Arriving at ESO Paranal is like crossing an international border. We had to wait for a formidable metal gate to open, show our passports, collect electronic ID badges, and stay in a gatehouse until someone in

3. *Europe's Quest for the Universe*, L. Woltjer (EDP Sciences 2006), p16.

authority came to collect us. The astronomers' residence, which featured in the closing portion of the James Bond movie *Quantum of Solace*, has many well-appointed bedrooms, meeting rooms, a large cafeteria, and even a swimming pool. The residence was full at the time, owing to the buzz of activity associated with the building of ESO's next big thing—the Extremely Large Telescope (ELT), a 39-metre segmented-mirror telescope under construction on a nearby peak, Cerro Armazones.

Witnessing Cerro Paranal and the enormity of the huge enclosures containing each of four 8-metre telescopes is almost akin to a religious experience (plate 47). You cannot really contemplate the scale of construction until you're standing close by on the mountain (whose original peak was flattened with dynamite) and have stepped inside each telescopic enclosure and gasped at the enormity of each so-called unit telescope. Individual indigenous Mapuche names were later chosen for all four of these giant telescopes through an essay contest among local schoolchildren.[4]

After leaving Caltech, I became an ESO employee based at its European headquarters in Garching in September 2015. Since ESO has a mandatory retirement at age 67, this was a brief 2-year appointment as a "senior visiting scientist," at the kind invitation of the director-general, another Dutchman, Tim de Zeeuw. This was an exciting time to arrive at ESO, since the largest ongoing project at the time was the construction of the 39-metre ELT (plate 48).

The ESO director-general is probably the most powerful person in world astronomy. The position became vacant in 2006, and I was encouraged to apply by Bob Williams, Matt Mountain (respectively former and, at the time, incumbent director of the Space Telescope Science Institute), and Alvio Renzini (a distinguished Italian astronomer who had worked very effectively as director of science at ESO for some years). My wife Barbara was already hoping to return to Europe so I considered it carefully but, in the end, I was put off when I met Ian Corbett (introduced in chapter 5), who was working at ESO as a senior

---

4. Their chosen unit telescope names are *Antu* (Sun), *Kueyen* (Moon), *Melipal* (Southern Cross), and *Yepun* (Evening Star).

administrator with the outgoing director-general, Catherine Cesarsky. We had breakfast in Prague at the time of the triennial general assembly of the International Astronomical Union in 2006. Although Corbett encouraged me to apply, the picture he sketched, whereby I would be spending all my time in committee meetings dealing with complex international agreements and financial planning, dampened my enthusiasm for applying. I would clearly have no time for my own science. Aged 56, I believed I could still enjoy a decade or more of making scientific discoveries, and I was loath to lose momentum. The position was first offered to Rolf-Peter Kudritzki, a delightful extrovert German who always wore colourful Hawaiian shirts (appropriately he became director of the Institute for Astronomy in Hawaii when I turned down that post in 1999).[5] In discussions with the council, ESO's powerful oversight body composed of both senior astronomers and financial administrators from the member nations, Kudritzki failed to negotiate resources to maintain an active research programme. So the job then went to De Zeeuw.

De Zeeuw was a Leiden professor when I was at Cambridge, and I knew him well. A very capable astronomer who specialised in stellar dynamics, he was ambitious and hard-working and successfully got the ELT project started. He worked hard to get me to ESO even though its council was apparently not that enthusiastic. One problem was where within the organisation I would best fit. The fact that I had negotiated a professorship back at UCL from 2017 and had secured a large European Research Council grant to support me, plus the fact that my ESO contract would be a temporary arrangement, finally won the day.

Arriving at ESO headquarters from Caltech was a bit like landing on another planet. I was sent a "New Staff Information Pack" (72 pages) and a "Staff Regulations" document (163 pages). As an employee of a treaty organisation, I would be given special residence status in Germany (termed *Sonderausweis*), my salary would not be taxed, and I would enjoy immunity from various jurisdictions, although the

---

5. Remarkably, his fascination with Hawaiian shirts predated his appointment as director of the Institute for Astronomy in Honolulu.

documents warned that this did not include parking and speeding tickets! Apart from 2 years at the former Royal Greenwich Observatory (1983–1985), a government agency, I had by now worked in academia for over 40 years. The atmosphere in a university department is largely established by its young students. Undergraduates cram the corridors in between classes and come to your office unannounced with trivial questions and their personal problems. I enjoyed their company, given their youthful enthusiasm. Graduate students and postdocs are naturally more mature but equally exciting to have around as they represent the lifeblood of academic research.

ESO headquarters was not an academic institution. It was an administrative and operational centre with telescope and software engineers, instrument specialists of all kinds, and project managers. Including a large human resources group, and relocation experts, the ESO staff list totalled 600 employees, split between Garching and Santiago, Chile. There was a research unit led by a director of science, Rob Ivison, a tall Yorkshireman who was my boss and generously left me alone. This unit comprised ESO research fellows, early career researchers, some of whom had operational duties alongside their own research programmes. No undergraduates, no postgraduates, no teaching. Walking in the corridors I encountered serious, businesslike colleagues who nodded quietly or gently muttered the Bavarian greeting *Grüss Gott*. Although the working language was English, the ESO staff were truly international. During coffee breaks, French, Swedish, Italian, and German workers gathered with their national colleagues and conversed in their own languages.

Established in a spacious office at ESO in late August 2015, I began to consider how to make progress on probing more extensively into the reionisation era. I returned to the fundamental question of whether early galaxies were capable of transforming the hydrogen gas around them into constituent protons and electrons. And what was the composition of the gas within these galaxies? Presumably as we probed closer to cosmic dawn, there would be fewer heavy elements such as carbon and oxygen, given those atoms could be synthesised only in previous generations of stars that had exploded as supernovae. While the numerous photometric surveys of deep Hubble images, both in the deep fields and aided via gravitational lensing through foreground clusters, had located candidates to redshifts as high as $z \sim 10$, none was spectroscopically confirmed and we knew very little about their properties. The VLT had numerous first-rate scientific instruments, and I was eager to make use of them.

I was also concerned about the hypothesis expounded in our well-cited papers led by Brant Robertson that galaxies governed the reionisation process. Although we now had a reasonable guess of how many galaxies there might be out to redshifts of $z \sim 10$, to verify the hypothesis we needed to know *how many ionising photons* were produced in an average galaxy. A few quasars had by now also been detected in the reionisation era. As introduced in chapter 3, these are rare, spectacularly luminous galaxies with supermassive black holes in their centres. As material falls under gravity into the dark abyss of their black holes, more-energetic radiation could easily escape and contribute to the reionisation process. A provocative article by an American astronomer of Italian origin, Piero Madau, and his colleague emerged that summer entitled "Cosmic Reionization after Planck: Could Quasars Do It All?" How could this new idea be addressed?

Throughout my career I had been an advocate of spectroscopy, fund-raising for and overseeing the construction of several spectrographs. While Hubble's striking deep images had clearly pushed back the horizons and revealed galaxies to the earliest times, to answer the scientific questions we now faced we needed spectra. However, if the Lyman-alpha emission line so successfully exploited by Dan Stark and others to redshifts of $z \sim 7$ would be increasingly undetectable in the reionisation era owing to scattering by the hydrogen clouds in the intergalactic medium, we needed to consider other diagnostic lines to address the chemical composition of the gas in galaxies, the production rate of ionising photons, and whether the radiation originated from gas clouds heated by starlight or emerged directly from material falling into a black hole.

ESO offered a remarkable spectrograph called X-shooter on one of the VLT's unit telescopes. This was capable of covering almost the entire optical and near-infrared wavelength range accessible from the ground in a single exposure. Meanwhile, Keck had augmented NIRSPEC, which Dan Stark, Matt Schenker, and I had championed earlier (chapter 8), with a new *multi-object* near-infrared spectrograph, MOSFIRE (the Multi-Object Spectrograph for Near-Infrared Exploration), built by Ian McLean and Chuck Steidel; this could tackle up to 20 sources at a time. However, neither NIRSPEC nor MOSFIRE could simultaneously study a very wide wavelength range. The ability to search for several emission lines other than just Lyman-alpha was a key advantage of X-shooter.

In pushing further into the reionisation era with the VLT, one question was how to select these targets. Until this point, virtually all spectroscopy had chosen candidates from deep, multicolour Hubble images, either gravitationally lensed examples seen through foreground clusters or sources in blank fields such as the HUDF. Many of Hubble's deepest fields were now also being imaged in the infrared by the Spitzer Space Telescope. Named after Lyman Spitzer, the pioneering Princeton astrophysicist introduced in chapter 6 who played a key role in championing space astrophysics, this modest 85-centimetre-aperture telescope was often in the scientific shadow of its larger sister, the Hubble Space Telescope. Although its conception can be traced back to a proposed Space Infrared Telescope Facility (SIRTF) in 1969, it became part of a NASA Great Observatories vision in 1985 and thereafter was led by scientists and engineers at the Jet Propulsion Laboratory, which is managed by Caltech in Pasadena. After many hiccups, including cancellation, revival, and descopes,[6] it was launched into a solar orbit in 2003 soon after I arrived at Caltech. As we will see, the Spitzer Space Telescope played a decisive role in probing the reionisation era.

As primarily an optical telescope, Hubble can penetrate only a small portion of the near-infrared waveband. For the large redshift of an early galaxy, this corresponds to probing the *ultraviolet radiation* that left that source. Spitzer, on the other hand, is a fully fledged infrared observatory

---

6. The remarkable history and scientific achievements of the Spitzer Space Telescope are described from first-hand experience by Michael Werner and Peter Eisenhardt in their lavishly illustrated book *More Things in the Heavens* (Princeton University Press 2019).

that extends the wavelength coverage significantly, not just beyond the range of Hubble but, for faint sources, beyond that which can be efficiently covered by large ground-based telescopes. As an infrared facility, Spitzer samples the light that left an early galaxy in the *optical region*. After a century of studies with optical telescopes, astronomers have learnt how to deduce many physical properties of galaxies from features in this familiar wavelength range. Clearly, the combination of Hubble and Spitzer would be very powerful, not only for learning more about early galaxies but also in selecting key targets for ground-based spectroscopy.

In addition to finding my feet at ESO, I was also slowly building up my new research team at UCL. During my first visits to the university, I met Guido Roberts-Borsani, a first-year PhD student who had spent the previous summer doing a research project with Rychard Bouwens at Leiden University. With an Italian mother and an English father, and being a Belgian national, he could hardly be more international! We got on really well, not least because he was a big fan of rugby union and Premier League football. Under Bouwens's supervision, Roberts-Borsani matched Hubble and Spitzer images and searched for those galaxies with estimated redshifts greater than $z \sim 7$ that also had a boosted signal, or "excess," in one of the Spitzer bands compared to that in the other.

With its modest aperture, Spitzer was not able to undertake infrared spectroscopy of faint galaxies, but several earlier workers had suggested an excess signal seen in one or other of its bands might arise from intense emission lines. The idea was first demonstrated by Hyunjin Shim and Ranga-Ram Chary at the Spitzer Science Center in Pasadena, who located a Spitzer excess attributed to the Balmer H-alpha line of hydrogen for galaxies in the redshift range $z \sim 3.8$ to $z \sim 5.0$.[7] Dan Stark, Matt Schenker, and I undertook a similar exercise using Spitzer data for

7. The Balmer H-alpha line of hydrogen is emitted in the red region of the optical spectrum and arises from a different change of electron energy than the Lyman-alpha line (discussed earlier), which originates in the ultraviolet spectral region. Although H-alpha is a prominent feature in low-redshift galaxies, and a valuable diagnostic of the rate of star formation in galaxies, it is unfortunately redshifted beyond the reach of ground-based telescopes at redshifts $z > 3$.

Stark's extensive DEIMOS redshift survey discussed in chapter 8. Using the technique at higher redshifts was first undertaken by Ivo Labbé and Renske Smit, both at the time in Leiden. Smit showed that a significant fraction of galaxies in the redshift range $z \sim 6.6$ to $z \sim 7.0$ had a Spitzer excess that was most naturally explained via the presence of surprisingly intense oxygen emission lines. She also demonstrated how attributing this excess to a known emission line can be valuable in confirming the redshifts of candidates without spectroscopic data. Smit's discovery has also been important in defining how to locate suitable *lower redshift analogues* of galaxies in the reionisation era. As these are closer and brighter, such analogues can be studied in much greater detail, thereby shedding light on how galaxies may drive the reionisation process.

More than anyone, Ivo Labbé deserves the credit for fully exploiting the Spitzer Space Telescope for studies of galaxies in the reionisation era. Not only did he secure observing time to undertake the deepest observations, but he developed sophisticated techniques to match the resulting data with associated images taken with the Hubble Space Telescope. I greatly admired Labbé's talents and expertise and was sure I could learn a lot from talking to him. Sadly, we rarely met in person.

There are two infrared Spitzer bands, at wavelengths of 3.6 and 4.5 microns. The oxygen emission line Smit proposed in the redshift range $z \sim 6.6$ to $z \sim 7.0$ causes a "Spitzer excess" in the 3.6-micron band. The same line would produce an excess in the 4.5-micron band for higher redshift galaxies between $z \sim 7$ and $z \sim 9$. Intriguingly, Roberts-Borsani presented the case for four such $z > 7$ Spitzer-excess galaxies (plate 49). Although the same oxygen emission lines were seen in nearby star-forming galaxies, as in Renske Smit's earlier work, the excess signal at high redshifts implied the lines were much stronger. Roberts-Borsani's models indicated his $z > 7$ galaxies might be very young, conceivably only 5 million years old. Were these galaxies being observed at a special moment in their early history?

Pascal Oesch (chapter 6), a co-author of Roberts-Borsani's article, was first to follow up the latter's targets using Keck's MOSFIRE spectrograph. He secured a redshift $z = 7.73$ for EGS-zs8-1 in April 2014, consistent with its colour-estimated redshift of 7.9 ± 0.3 (upper right in plate 49). This was a new redshift record, surpassing one at $z = 7.508$ by Steve Finkelstein (University of Texas) in 2013, which in turn eclipsed the Japanese record at $z = 7.213$ (chapter 8). Surprisingly, Oesch's redshift was based on the Lyman-alpha line of hydrogen, which was clearly visible after only a 4-hour exposure.

The international race to follow up Roberts-Borsani's other sources was on, and my colleagues and I were not far behind. During my last 6 months at Caltech, Israeli research fellow Adi Zitrin and I used Keck's MOSFIRE to target the most distant candidate in Roberts-Borsani's list, EGSY8p7, at a colour-estimated redshift of 8.6 ± 0.3 (upper left in plate 49). Sure enough, we secured a redshift $z = 8.683$, again based upon Lyman-alpha emission, in an exposure of only 4.3 hours. Zitrin is one of the fastest people I have ever worked with. The relevant observations were taken on the night of June 10, 2015, yet the article he led was submitted for publication less than a month later on July 9, appearing in print with a press release as I left Caltech for ESO on August 28. As a result of Zitrin's efficiency, Oesch's record held for only a few months.

While it was exciting to break the redshift $z = 8$ barrier, the new results presented a conundrum. All four of Roberts-Borsani's Spitzer-excess candidates were eventually confirmed beyond redshift $z \sim 7$ through clear detections of Lyman-alpha emission.[8] How could this be, given that these sources, deep in the reionisation era, should be surrounded by clouds of atomic hydrogen capable of intercepting and scattering any Lyman-alpha photons? The puzzle was particularly acute for Adi Zitrin's object at a redshift $z = 8.68$, where, according to the Planck team, reionisation was still at an early stage. Matt Schenker, Dan Stark, and I had pounded away with Keck on many other candidates supposedly beyond a redshift $z \sim 7$ and not seen Lyman-alpha emission in any of them. Other groups had followed, confirming the absence of the line in similarly selected sources. The only differences between these much more numerous earlier targets and the four selected by Roberts-Borsani were their

8. EGS-zs8-2 was confirmed at a redshift $z = 7.477$ from Oesch's observing run in April 2014, and the result was included in an update to Roberts-Borsani's published article. COSY was confirmed at a redshift $z = 7.154$ during our Keck run in November 2015 in collaboration with Dan Stark and Adi Zitrin.

brightness and the additional constraint of a Spitzer excess. Whichever way you looked at it, there seemed to be something special about Roberts-Borsani's sources that enabled Lyman-alpha to be visible.

Following Brant Robertson's papers, the popular view in the community was that reionisation was driven by feeble, low-luminosity galaxies. They were the most numerous and thus produced the lion's share of ionising photons. But suppose rarer, luminous objects like those in Roberts-Borsani's list were somehow more energetic and had a head start, creating their own ionised bubbles so that Lyman-alpha could readily escape any scattering by atomic hydrogen? Perhaps these more massive galaxies had enjoyed more time to grow massive black holes in their centres and that gave them an extra ionising capability? To test this idea, we needed to study these Spitzer-excess sources in more detail.

My European grant at UCL was ambitiously entitled "First Light" (to my dismay, someone else had already grabbed the title "Cosmic Dawn") and provided salaries for several postdocs and students as well as travel funds to go observing, attend conferences, and shuttle back and forth between Munich and London on a regular basis. My first postdoc hire was an imposing barrel-chested Frenchman, Nicolas Laporte—another rugby fan. Laporte did his PhD in Toulouse and so, naturally, was an expert on gravitational lensing. He then moved to the Catholic University in Santiago, where he gained valuable observing experience using many Chilean telescopes. Recommended by Jean-Paul Kneib, Laporte became the primary high-redshift observer in my UCL group for the next 3 to 4 years.

Laporte and I wondered whether we could use X-shooter to figure out what was going on with Roberts-Borsani's Spitzer-excess sources. Could we somehow determine if they contained massive black holes into which infalling material would generate powerful ionising radiation? Dan Stark had been working with a theoretical astronomer in Paris, Stéphane Charlot, and his students Anna Feltre and Julia Gutkin, who could predict which gaseous spectrum lines might distinguish whether early galaxies were ionising their surroundings via massive black holes as opposed to starlight.

The radiation from a star is determined largely by its surface temperature. The hotter the star, the greater the proportion is emitted in the ultraviolet region, where photons have higher energies. Such photons can excite the helium, carbon, nitrogen, and oxygen atoms in the gas in the interstellar medium between stars into certain energy states that reflect the strength of this stellar radiation. Given the mix of stellar types and their respective temperatures, Charlot's team was able to calculate the range of spectrum lines that should be observed from the hot, excited gas within a galaxy.

However, the radiation emerging from gas falling into a massive black hole at the centre of a galaxy is qualitatively different. As the infalling gas is fully ionised, free electrons can be accelerated around compressed magnetic fields, releasing more powerful radiation, which can accordingly excite the gas into higher energy states. Since not all massive black holes are surrounded by infalling gas, the term "active galactic nucleus" (AGN) is sometime used to distinguish ones capable of generating this intense radiation as opposed to those where the black hole is present but effectively dormant. Charlot's team could also predict which spectrum lines would diagnose the presence of AGNs in our growing sample of distant galaxies.

This new campaign was very demanding observationally. The most energetic lines of helium, carbon, nitrogen, and oxygen are often weak because the radiation emanating from the central black hole, while intrinsically powerful, still represents only a small fraction of that emerging from starlight across the rest of the galaxy. Nicolas Laporte led the charge with X-shooter, undertaking heroic exposures of up to 12 hours. Even after these long exposures, we were often looking at marginal smudges on our computer screens, wondering whether anyone would believe our results. At Arizona, Dan Stark and his student Ramesh Mainali used MOSFIRE on Keck to undertake a similar campaign. Collectively, we uncovered promising evidence of AGNs in several systems. Specifically, for Roberts-Borsani's four Spitzer-excess galaxies at redshifts greater than $z \sim 7$ (plate 49), two showed a very energetic line of nitrogen, as well as, in one case, velocity-broadened Lyman-alpha, both hallmarks of an AGN. In total, the Keck and VLT surveys found evidence of an unusually strong radiation field in almost half the spectroscopically confirmed galaxies beyond redshift $z \sim 7$ known at the time. Although this was promising progress, indicating the likely development of massive black holes in the early universe, the next logical step—measuring the black hole masses—was simply too difficult. After a considerable effort, we had simply strengthened the view that early galaxies, especially the massive ones and those with Spitzer excesses, generated large amounts of ionising radiation, perhaps as a result of growing black holes.

Spectroscopy had a lot of catching up to do, given a flurry of potential distant candidates arriving from deep Hubble and Spitzer Space Telescope imaging. Unfortunately, many candidates were simply too faint even for the Keck and VLT facilities. The nature of the most distant candidate in the HUDF, UDFj-39546284 (chapter 8) remained unclear, despite a spectroscopic attempt using Keck by Peter Capak and colleagues in Pasadena. A number of gravitationally lensed candidates thought to lie beyond a redshift $z \sim 9$ were also the subject of much debate at conferences.

Let's go back for a moment to review how spectrographs work. The earliest spectrographs employed a *prism*. Although Isaac Newton is often considered the father of the subject following his treatise *Opticks* (1704), the dispersing power of a glass prism was familiar to natural philosophers centuries earlier. However, the spectrographs on all the telescopes I've discussed thus far employ a *diffraction grating*, a reflecting surface or transmitting optical component ruled with a series of narrowly separated ridges. Because of the wave-like nature of light, constructive interference between the light reflected from the adjacent ridges disperses the light into a spectrum on the detector. A grating can be very finely ruled, offering superior spectral resolution (the ability to separate features close in wavelength) compared with a prism. An entrance slit or multi-slit aperture mask ensures only the light from the required target(s) is observed.

The WFC3 onboard Hubble is equipped with a *grism*. This is a combination of a prism and grating arranged so that, for a selected wavelength, the light passes straight through. When this grism is inserted into the light beam it delivers a spectrum for every single galaxy in the field of view of a Hubble image. While this may seem highly advantageous in terms of multiplexing efficiency, the problem is that the spectrum of every source becomes a thin stripe on the detector that often overlaps with ones arising from neighbouring sources (plate 50). This spectral contamination is a big problem in fields containing many sources, such as lensing clusters. It can sometimes be mitigated by rotating the grism through 90 degrees and taking a second exposure in the hope that there will be a better (or, at least, different) geometrical arrangement of overlapping spectra. I have never been a big fan of grisms for a further reason. Clearly, to minimise this overlapping issue the spectra should be as short as possible on the detector, and this means a low spectral resolution; sharp emission lines such as Lyman-alpha become blurred and lie undetected unless they are very intense.

Despite these challenges, many teams locating gravitationally lensed sources behind foreground clusters were attempting to use Hubble's WFC3 grism to determine spectroscopic redshifts for candidates with redshifts as high as $z \sim 10$ to 11. For example, Dan Coe at the Space Telescope Science Institute (STScI) located a triply imaged source, MACS0647-JD, seen behind the foreground cluster MACS0647 + 7015 with a colour-based redshift of $z \sim 10.7 \pm 0.5$ (plate 50, *top*). As this source had no convincing Spitzer detections, it was detected in only two out of over a dozen Hubble colour filters. Although the geometrical arrangement of three multiple images might give some idea of the distance of the source (see chapter 7), unfortunately Adi Zitrin's models for the cluster did not definitely rule out an intermediate redshift source. Subsequently, Nor Pirzkal, also at STScI, secured a grism spectrum that showed no convincing signal, although he claimed the absence of any strong lines in his spectrum ruled out a low-redshift source. A similar story followed the discovery of a further triply imaged source, A2744-JD1, where Zitrin predicted a record-breaking redshift of $z = 9.8$. Again, there were no Spitzer detections, but this time the lensing model provided a tighter geometrical constraint (plate 50, *bottom*). Sadly, this source was too faint for useful spectroscopy.

These various Hubble surveys for gravitationally lensed galaxies at high redshift were originally proposed with the claim that they would deliver highly magnified sources bright enough for detailed spectroscopy. It was argued by many that such lensing surveys would offer a unique advantage compared with chasing similar sources in the HUDF and CANDELS surveys, which would be too faint to follow up. The most prominent lensing campaign was the Hubble Frontier Fields programme, led by Jennifer Lotz, then at STScI, which undertook extremely deep imaging of six carefully selected clusters. Yet, frankly, as regards providing detailed spectroscopy of early galaxies, the returns were disappointing. Highly magnified sources are extremely rare, as Jean-Paul Kneib and I had found 10 years earlier (chapter 7). While large numbers of early galaxies were catalogued as a result of these lensing campaigns, most are only modestly magnified and too faint for spectroscopic study.

As if to rub salt in the wounds of lensing aficionados who failed to spectroscopically confirm their candidates, the first truly interesting source confirmed beyond a redshift $z \sim 10$ came not from a lensing campaign but from the CANDELS survey. In 2014, Pascal Oesch and collaborators continued their census of redshift $z \sim 9$–10 candidates, incorporating Spitzer to provide additional evidence for their high-redshift nature. An early WFC3 grism spectrum of one of them, labelled GN-z10-1 at the time, did not show strong features, but Oesch successfully proposed an integration three times longer, optimised for studying GN-z10-1. In March 2016, Oesch and the full armoury of the Illingworth team announced the discovery of the (renamed) GN-z11, in an article entitled "A Remarkably Luminous Galaxy at $z = 11.1$ Measured with Hubble Space Telescope Grism Spectroscopy." Within the space of only 7 months, the record Adi Zitrin and I had established at a redshift $z = 8.683$ had been significantly surpassed!

Oesch's spectrum was taken at two orientations with the WFC3 grism but did not reveal any emission lines. However, he presented a sharp break in signal indicative of hydrogen absorption at a redshift $z = 11.1 \pm 0.1$. This break was the same feature originally used to select the source as a potential high-redshift candidate from the Hubble multi-band imaging;[9] the only difference was it was recovered here with a

9. The discontinuity due to hydrogen absorption was also detected for the gravitationally lensed $z \sim 7$ galaxy in the cluster field Abell 2218 by Jean-Paul Kneib, Mike Santos, and I discussed in chapter 7.

grism spectrum rather than by examining the galaxy's colours. Yet the numbers didn't quite match, as the need to change the name of the source made abundantly clear. The Hubble multi-band imaging published in 2014 originally located this hydrogen break at a redshift of $z = 10.2 \pm 0.4$. Indeed, the imaging data included a clear photometric detection seemingly inconsistent with a break at the higher redshift of $z = 11.1$ (plate 51). It seemed to me that GN-z11 remained a convincing redshift $z \sim 10$ candidate, but the new grism data was too marginal to justify adopting the higher redshift. A further attempt at clinching the redshift of this enigmatic source using Keck's MOSFIRE was undertaken by a large team led by Linhua Jiang at the Kavli Institute in Beijing. After a 5-hour exposure, the team claimed detection of two emission lines of carbon and one of oxygen consistent with a redshift $z = 10.957$. Of these three lines, two are quite weak, and the claimed spectroscopic redshift is still a little difficult to reconcile with the photometry.

Regardless of the above uncertainties, which serve to emphasise the well-known challenges of undertaking spectroscopy at the frontier, there is one further puzzle about GN-z11: it is extremely bright. In fact, it is brighter than most galaxies at redshifts $z \sim 6$ to 8. There was no evidence it was gravitationally lensed and magnified by a foreground system. A redshift $z \sim 11$ corresponds to a time only 400 million years after the Big Bang, when the universe was less than 3% of its present age. According to the Planck results (chapter 8), reionisation was just beginning. How could such a luminous galaxy already be present? Both Oesch and Jiang and their colleagues were equally puzzled, claiming that most likely the galaxy was surprisingly young.

The Planck team studying the cosmic microwave background did seem overconfident that cosmic reionisation began at redshift $z \sim 12$ and ended at a redshift $z \sim 6$. More-detailed simulations of the Planck results in the context of the accepted "cold dark matter" model of structure formation suggested there could be a tail of star-forming galaxies extending out to redshift $z \sim 20$, when the universe was less than 200 million years old. Sadly, neither Hubble nor ground-based telescopes could probe beyond a redshift $z \sim 11$ because the very sharp drop due to hydrogen absorption that Oesch and co-workers detected for GN-z11, and essential for selecting very early objects, would move to longer wavelengths beyond the range of their instruments. However, if you cannot witness the birth of an early galaxy directly, the next best thing would be to measure the ages of the most distant ones you can study. For example, if you could determine that a redshift $z \sim 8$–9 galaxy seen 600 million years after the Big Bang was already 400 million years old, it would imply the galaxy formed only 200 million years after the Big Bang, corresponding to a redshift $z \sim 19$, well beyond the reach of current facilities.

So how might one estimate the age of an early galaxy? Stars are born with a wide range of masses. The most massive ones are 50–100 times the mass of the sun and very hot. They burn their nuclear fuel rapidly and soon explode as supernovae. As the galaxy ages, stars of ever-decreasing mass evolve or die and thus no longer contribute to the overall population. By contrast, lower mass stars are cooler, burn less brightly, and live longer. Moreover, their cooler atmospheres are composed primarily of atomic hydrogen, which produces a characteristic absorption feature called the "Balmer break" in the blue spectral region.[10] The strength of such a feature in the spectrum of a galaxy would indicate the extent to which stars above a certain mass had evolved or died and hence indicate the age of the stellar population.

At a redshift $z \sim 8$–10, the Balmer break shifts into the infrared, and its signature is detectable only with the Spitzer Space Telescope. Earlier, I introduced the key role of this modest space telescope in finding Guido Roberts-Borsani's enigmatic redshift $z \sim 7$–8 galaxies with a "Spitzer excess," thought to arise from intense oxygen emission. Spitzer photometry can likewise be used to search for a Balmer break and hence to "age-date" early galaxies and pinpoint when cosmic dawn occurred.

In 2012 Wei Zheng, a postdoctoral researcher at Johns Hopkins University in Baltimore, discovered a faint gravitationally lensed source viewed through the MACS1149 + 2223 cluster of galaxies. Using both Hubble and Spitzer photometry for this galaxy, dubbed MACS1149-JD1,

---

10. This break arises from a sequence of atomic absorption lines of hydrogen from its second energy level to higher states, which cluster in the blue spectral region.

Zheng and collaborators estimated a colour-based redshift of $z = 9.6 \pm 0.2$ and noted a "Spitzer excess," which they interpreted as a Balmer break consistent with a galaxy already 200 million years old at that early epoch. They claimed it could have formed as early as a redshift $z \sim 14$.

The key uncertainty in this pioneering study was the possibility of contamination of the Spitzer band by the intense oxygen emission lines inferred in the lower redshift sources studied by Guido Roberts-Borsani and Renske Smit. Those oxygen lines can occupy the relevant Spitzer band at redshifts up to $z = 9.0$, beyond which they shift out of the band and cannot contaminate it. So everything hinged on the accuracy of Zheng's colour-based redshift. If the redshift of MACS1149-JD1 was $z = 9.0$ or less, the Spitzer excess might be due to oxygen emission lines and the age of the galaxy could be much younger. In short, for Spitzer to be useful in age-dating early galaxies, these galaxies needed to be spectroscopically confirmed to lie at a redshift greater than $z = 9.0$.

Recognising this important requirement, Austin Hoag, a graduate student at the University of California, Davis, working under the supervision of Maruša Bradač, an energetic and athletic Slovenian, secured a deep WFC3 grism spectrum of JD1 (as this interesting source soon became known) and claimed to see a break due to hydrogen absorption (as done by Oesch for GN-z11) at a redshift of $z = 9.5 \pm 0.5$. Normally, a spectroscopic redshift is more accurate than one based on colours, but in this case the grism data suffered serious contamination from light from nearby sources, rendering the result quite uncertain. Although a heroic attempt, it did not remove the possibility of a lower redshift and younger source.

Nicolas Laporte, Guido Roberts-Borsani, and I set out to Chile in early 2018 to settle the matter with X-shooter on the VLT (plate 52). Although I had observed at Cerro Paranal when the first 8-metre unit telescope opened in 1999, this was my first observing campaign as an ESO staff member. Consistent with ESO's official policy, visiting astronomers must sit passively, commenting on decisions and primarily focusing on analysing their incoming data. All instrument and telescope control is strictly the responsibility of ESO staff. This was humiliating for someone with four decades of hands-on observing experience, but sadly it is increasingly commonplace at large observatories. Such a detached mode of observing eliminates the romance and excitement of discovery. More importantly, it means young students and postdocs are not easily trained to learn the responsibilities associated with getting quality data. I write this without in any way denigrating the quality and professionalism of the support staff at the VLT; their experience and enthusiasm certainly matched those I witnessed at Keck. They were just following ESO's rules.

Laporte, Roberts-Borsani, and I were frustrated during our time at Cerro Paranal. We were trying to address a pretty important question: When did galaxies first form? On one night, although there was no cloud, the humidity hovered at around 50% all night, precluding any observations. The observatory staff retired to watch movies. I pointed out that, at Keck, telescope operators would routinely continue to observe at a humidity level of 80%, but this didn't impress the locals. Eventually I was told where to stuff my complaints. After losing much of the scheduled time but nonetheless seeing a promising signal, I appealed to my boss at ESO, Rob Ivison, who happened to be visiting the office in Santiago, where I was giving a seminar after the observing run. To his credit, Ivison very kindly provided additional observing time when, to our delight, we finally clinched a redshift $z = 9.11$ for JD1 based on Lyman-alpha emission.

Back in London, we were over the moon. We now had a spectroscopic redshift proving beyond doubt that the Spitzer-excess signal in JD1 was due to a Balmer break consistent with a mature galaxy whose age we could estimate. Although this idea was championed by Wei Zheng, our redshift differed from his earlier estimates and was much more precise. Soon after, however, Laporte came into my office with a long face and said he'd noticed that a Japanese team was simultaneously studying JD1 with the Atacama Large Millimetre Array (ALMA), the powerful millimetre interferometer, also in Chile. The Japanese were apparently scanning the redshift range $z \sim 8.8$ to 9.5 for a far-infrared oxygen emission line. With characteristic efficiency, Laporte had already downloaded the first portion of their data, which scanned the higher redshift range first proposed by Hoag and Bradač and by Zheng.

Consistent with our VLT redshift for JD1, he found no oxygen emission in their data. Frustratingly, however, although the ALMA data in the redshift range encompassing our VLT redshift had been taken, they had not yet been made public. Only the Japanese team had access to them. Quite likely they had by now identified the redshift of JD1 independently and thus might scoop our result by racing to publication. We agreed that Laporte would write to the leader of the Japanese team, Akio Inoue, at the time a professor at Osaka Sangyo University, with a carefully crafted email predicting they would find oxygen emission in their recently acquired proprietary data. Would he confirm this so we could verify our own VLT redshift?

Patience is not one of Laporte's virtues; he pestered me every morning when I got to the office. He repeatedly asked, "Why don't the Japanese respond?" To placate this large Frenchman, I suggested that silence was a good omen. I imagined they indeed had a redshift identical to ours with ALMA but were wondering how to respond, as otherwise they would have delighted in telling us the opposite. Sure enough, a week later, they confirmed our redshift (plate 53, *right panels*), and, following extensive and sometimes delicate discussions over the internet, we agreed to submit a joint paper led by Inoue's postdoc Takuya Hashimoto and Laporte for the journal *Nature*.[11] The article appeared in May 2018 with accompanying press releases from UCL, ESO, and Japan.

The *Nature* paper was entitled "The Onset of Star Formation 250 Million Years after the Big Bang" and had 24 authors, including Wei Zheng, who first considered the prospects of age-dating this unique source. The spectroscopic redshift $z = 9.11$, unusually confirmed independently by two telescopes, eliminated any possibility that the Spitzer excess arose from intense optical oxygen emission, as was the case for Roberts-Borsani's lower redshift sources (plate 53, *lower left*). This enabled co-author Ken Mawatari to use newly available deeper Spitzer data to refine the age of

11. The entire drama of the UCL-Japanese interchange is retold in a Japanese NHK television documentary, where Akio Inoue and Takuya Hashimoto are filmed reading Laporte's email and discussing how to react. They re-enact a discussion where they decided to inform us that they have much better data than they imagine we have!

JD1. His best-fit model had a dominant old stellar population that formed at a redshift $z = 15 \pm 2$ with an additional secondary burst of star formation just before a redshift $z = 9.11$ (plate 53, *lower left*, black and blue curves). Although it is always wise to be cautious on inferring too much from one source, MACS1149-JD1 provided an encouraging first glimpse of when the earliest galaxies emerged from darkness.

The age-dependent Balmer break, unambiguously identifiable only through data taken with the Spitzer Space Telescope for galaxies beyond a redshift $z = 9.0$, encouraged us to search for further $z > 9$ galaxies to see if our estimate of when JD1 switched on its starlight was representative of other early systems. Laporte and I located a further five candidates whose colours suggested they lay beyond a redshift $z \sim 9$, none of which yet had spectroscopic redshifts. With a UCL graduate student, Romain Meyer, Laporte and I put great effort into securing observing time on almost every possible large telescope to make progress. We teamed up with Brant Robertson and Guido Roberts-Borsani and, over a 3-year effort, secured spectra for all five sources using Keck, ESO's VLT, the southern Gemini 8-metre telescope, and the ALMA interferometer. We secured redshifts for three of the five galaxies, including a convincing redshift of $z = 9.28$ for a further lensed source, MACS0416-JD. Including a reanalysis of the earlier source MACS1149-JD1, previously studied with our Japanese colleagues, we estimated stellar ages for all six sources and refined our estimates of cosmic dawn (plate 54).

The range of ages we determined indicate that cosmic dawn was probably a gradual rather than an instantaneous event, consistent with occurring in the time window of 250 to 350 million years after the Big Bang, corresponding to a redshift range from $z \sim 16$ to $z \sim 13$. We also estimated how luminous our galaxies would have been at earlier times. This was done by modelling their observed colours in the context of the most likely history of how stars assembled and evolved in each galaxy. Although it is not possible to observe any celestial object at different times in its history, if our sources are representative of others at $z \sim 9$, this should give a reasonable indication of whether future facilities, such as the James Webb Space Telescope (JWST; see the epilogue), could directly witness primordial emerging from darkness at cosmic dawn. Whilst this kind of "backtracking" exercise is never completely free from assumptions, our calculations showed that similar sources would be detectable, close to the time of their birth, in exposure times of only 3 hours with JWST. More challenging spectroscopy would then be needed to confirm their redshifts.

And here the current story ends. From humble beginnings with the AAT, using a primitive fibre-fed spectrograph to chart the redshift distribution of galaxies with the simple goal of understanding galaxy evolution since a redshift $z \sim 1$, advances in technology and the ambitions of numerous astronomers have led to the construction of larger, more powerful telescopes on the ground armed with more-efficient instruments and improved detectors. In a partnership with both the Hubble and Spitzer Space Telescopes, the frontiers have been pushed back to ever-higher redshifts, reaching close to $z \sim 11$. As in all areas of science, there have been numerous premature claims, including several by myself and colleagues, but collectively observational astronomers have pieced together a picture of the evolution of galaxies and their role in transforming the hydrogen in intergalactic space back over 97% of cosmic history. The most recent development is an indirect glimpse of cosmic dawn, predicting when it probably occurred and how luminous similar sources emerging from the darkness might be. As so often in science, the best is yet to come. What might we find at cosmic dawn? Numerical simulations suggest the end of darkness is likely accompanied by the rapid collapse of gas clouds, fuelled by further material arriving along thin, wispy filaments. Young stellar systems flicker into existence and their massive stars soon explode as supernovae, enriching space with processed chemical elements. The entire event would be astonishingly lively to watch (plate 55).

Allan Sandage once told me that Edwin Hubble was not a great observer, which is why he needed to rely on the assistance of Milton Humason. However, Hubble clearly had a wonderful talent for majestic prose, matched, perhaps, only by George Ellery Hale's. It's a fitting conclusion to my story to use a famous quote from Hubble based on a series of lectures he gave at Yale University in 1935. It is as relevant today as it was nearly 90 years ago. "From our home on the Earth, we look out into the distance and strive to imagine the sort of world into which we were born. Today, we have reached far into space. Our immediate neighborhood we know rather intimately. But with increasing distance our knowledge fades, and fades rapidly. Eventually, we reach the dim boundary—the utmost limits of our telescopes. . . . The search will continue. The urge is older than history. It is not satisfied and will not be suppressed."[12]

12. From Our Home on the Earth, Edwin Hubble, *The Land*, vol 5 (1946), p145.

A Promising Future

It is a measure of the confident, ambitious spirit of American science that, in September 1993, only a few months after the successful repair of the Hubble Space Telescope, AURA and NASA initiated a study to consider "possible missions and programs for ultraviolet, optical and infrared astronomy in space for the first decades of the 21st century." One of the major recommendations of this study, conducted by a panel of 18 astronomers, including myself, was the proposed construction of a Next Generation Space Telescope (NGST), a 4-metre-aperture telescope optimised for the infrared spectral region. Although Riccardo Giacconi, director of STScI, had initiated a science workshop focusing on a similar theme organised by Garth Illingworth earlier in 1989, the discovery of HST's spherical aberration put further consideration of such a facility on hold. The AURA panel was chaired by Alan Dressler, a well-respected extragalactic astronomer at the Carnegie Observatories, who did a masterful job in writing the panel's report, entitled *HST and Beyond*, which was published in May 1996.[1] The scientific motivation for the proposed NGST is described in two chapters entitled "Visiting a Time when Galaxies Were Young" (i.e., the quest for cosmic dawn), and "The Search for Earth-like Planets and Life."

Dressler was initially a competitor of mine in studies of distant clusters of galaxies. For his PhD thesis at the University of California, Santa

---

1. *HST and Beyond: Exploration and the Search for Origins: A Vision for Ultraviolet-Optical-Infrared Space Astronomy*, Alan Dressler (AURA Publications 1996).

Cruz, he made definitive measurements of the "morphology-density" relation,[2] a trend whereby galaxies found in dense clusters are mostly red ellipticals or "lenticulars" devoid of star formation,[3] as opposed to the blue star-forming spirals more numerous in the less-dense regions commonly referred to as "the field." In other words, whether a galaxy is a spiral or an elliptical seems to depend on its local environment. There were two possible explanations for this observation. One was that, whereas ellipticals could form in over-dense regions, for some reason spirals could not. In this case, the morphology-density relation was established when galaxies formed. More intriguingly, however, another explanation is that spirals could somehow transform over time into ellipticals or lenticulars owing to physical processes unique to the dense environment of a cluster. These two contrasting hypotheses for the morphology-density relation became known as the "nature" and "nurture" explanations, respectively.

At about the same time as Dressler's thesis, Harvey Butcher at Kitt Peak and Gus Oemler at Yale University were examining the colours of galaxies in dense clusters up to redshifts $z \sim 0.5$ with a new intensified image tube camera. They claimed to find more blue, presumably star-forming, galaxies than those seen in nearby clusters. Their discovery supported the nurture hypothesis, and at the time this evolution of galaxies in clusters was regarded as an observational breakthrough as significant as the excess faint-galaxy counts discussed in chapter 4. However, this so-called "Butcher-Oemler" effect was controversial as it was unclear whether the blue galaxies were genuinely cluster members or simply spiral galaxies distributed along the line of sight in the lower density "field." To eliminate this alternative explanation meant getting redshifts for the blue galaxies.

While my colleagues and I were confirming the redshifts of these blue galaxies using the multi-object spectrographs we'd built for the AAT and WHT, Dressler and his US collaborators were doing the same

2. "Morphology" in this context refers to the visual classification of a galaxy (e.g., spiral, elliptical, or irregular).

3. Lenticular (or So) galaxies represent an intermediate class that has the disc-like structure of spirals but no star formation or spiral arms.

with instruments they'd built for the Hale 200-inch at Palomar. From both surveys, it seemed that distant clusters did indeed contain more blue member galaxies, and hence that the morphology-density relation was established relatively recently. Soon after HST was repaired in 1993, both competing teams recognised the benefits of combining forces to propose a large HST programme to examine the morphologies of these blue cluster galaxies. Were they really spiral galaxies that had somehow transformed their morphologies by the present day? The merged team, nicknamed the "Morphs," demonstrated that over the past 5 billion years spiral galaxies from the field were gravitationally drawn into clusters, and as they encountered the dense environments they were transformed by various physical processes into red ellipticals and lenticulars. This heralded a major advance in understanding the distribution of galaxy morphologies in the local universe.

Through the Morphs collaboration, I got to know Alan Dressler very well. A talented observer, he is eloquent and well organised, but also a determined individual with an occasional stubborn streak. Given these attributes and his stature in the US community, he was the ideal advocate for the NGST. Perhaps as a result of his recommendation, I was appointed as the only European member of the HST and Beyond panel, representing the interests of the European Space Agency (ESA), NASA's partner in HST.

To digress for a moment: As is well known, Europeans differ culturally from their American counterparts in many ways. One example is the European scientists' decision to take a decent summer holiday; don't expect any response from an Italian or French collaborator in the month of August! This insistence on a balanced lifestyle contrasts with the workaholic stereotype of the American scientist. Similar differences can also be found in the respective astronomical observatories. During my first observing run on the ESO 3.6-metre telescope at La Silla, I was astonished when the telescope operator announced he was taking a 45-minute break at midnight so he could retire to the canteen and enjoy a gourmet meal! I'd never encountered such disregard for precious observing time in Australia or the United States. Amusingly, I finally benefited from this European insistence on a comfortable lifestyle when

I learned that, as ESA's only member of the AURA panel, I was permitted to travel to its US-based meetings in business class.

The other members of Dressler's panel were distinguished US and Canadian astronomers. Dressler chaired our meetings effectively, and I learned a lot about US "astro-politics" in the process. At one panel meeting, a representative from a US aerospace company was scheduled to discuss possible designs for a 4-metre infrared space telescope. However, before beginning his presentation, he enquired which panel member was Dr Ellis. He explained that, for security reasons, such "aliens" were not permitted to receive his documents. Fortunately, a rebellious US panel member sitting beside me simply handed over his copy of the "forbidden" paperwork!

After the panel's recommendations were agreed, I was required to present the conclusions to senior ESA staff in Paris, who had paid my travel to participate in this exercise. An interesting section of Dressler's report emphasised American scientists' strong desire for public accountability. This was contained in a sympathetically written section of the report entitled "Sharing the Adventure: Inviting the Public Along." When I began talking about this aspect of our report during my ESA presentation, I was interrupted by the ESA director of science, Roger Bonnet, a wiry Frenchman of short stature, who interjected with some frustration, "Please, Dr Ellis, spare us this aspect of your presentation! We, in Europe, do not share this curious American desire to engage the general public."

Alan Dressler deserves much credit for propelling the NGST, which in 2002 was renamed the James Webb Space Telescope.[4] Key to early progress was the cordial relation Dressler established with NASA's administrator at the time, Daniel Goldin, who had begun a vigorous campaign to streamline the agency, with the slogan "cheaper, faster, and better." During a presentation at the American Astronomical Society's meeting in 1996, Goldin challenged Dressler in public to consider a

4. Breaking somewhat with tradition, NGST was named after James Webb (1906–1992), a NASA administrator who oversaw the agency's early manned space flight missions, rather than a famous physicist or astronomer, as was the case for each of the Great Observatories (Hubble, Spitzer, Chandra(sekhar), and Compton).

more powerful NGST with an 8-metre aperture, which, he suggested, might be constructed for only $500 million. Technical feasibility studies supervised by Hervey "Peter" Stockman (STScI) and John Mather (later Nobel Laureate in Physics 2006 and eventual JWST project scientist) based on work undertaken at NASA's Goddard Space Flight Center and by three independent aerospace companies, focused on an 8-metre telescope to be launched in 2008.[5] Following the precedent set by HST, STScI was announced as the eventual operating institution for JWST in 2000.

Following various reviews, in 2001 JWST was descoped to its present 6.5-metre-aperture telescope. Contracts for four scientific instruments were arranged over the next couple of years, but by 2004 the launch date had slipped to 2011. Perhaps through weary resignation, I failed to follow these delays with the same diligence as I had done for HST in the 1980s (chapter 5), but others kept the relevant record (figure 10.1). During 2005 the cost of the mission nearly doubled to $3.5 billion, with a launch date of 2013. In July 2011 the US House Appropriations Committee temporarily cancelled JWST in NASA's 2012 budget, owing to it being "billions of dollars over budget and plagued by poor management."[6] Considerable advocacy was required to rescue the project.

The Spitzer Space Telescope ceased operations in January 2020, bringing to an end nearly two decades of spectacular discoveries across all areas of astronomy. I was invited to summarise its extragalactic achievements at a major conference held at Caltech a month later. As discussed in chapter 9, this remarkable 85-centimetre space telescope was crucial in age-dating early galaxies and gave a first meaningful indication of when cosmic dawn occurred. This was my final visit to the United States before Covid-19 curtailed most international travel, so I was very fortunate to have had the opportunity to visit Northrop Grumman's facility in El Segundo, Los Angeles, on that occasion, where the near-complete JWST was in full view. Although it was nearly

5. *The Next Generation Space Telescope: Visiting a Time when Galaxies Were Young*, H. S. Stockman (ed.) (AURA Publications 1997).

6. Nasa Fights to Save the James Webb Space Telescope from the Axe, Robin McKie, *Guardian* (London) (July 9, 2011).

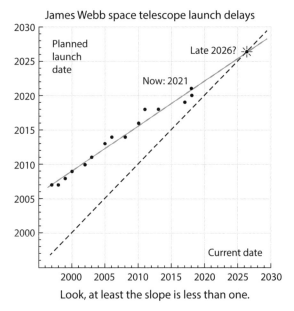

FIGURE 10.1. Forever slipping: continual delays to the predicted launch date of JWST from the late 1990s to 2017, with a (fortunately) pessimistic view of the eventual outcome.

24 years since my time on Dressler's HST and Beyond panel, witnessing the realisation of our primary recommendation was an inspirational moment (plate 56, *top*).

In addition to its impressive size, JWST has many innovative features. The primary mirror has an aperture of 6.5 metres, which gives it 6.25 times the light-gathering power of HST. A monolithic mirror of this size cannot be launched by any current space vehicle, and so, following Jerry Nelson's pioneering approach for the Keck telescopes, JWST's mirror (the responsibility of Ball Aerospace) comprises 18 hexagonal segments that can be folded into a more compact form to fit within the launch payload of ESA's *Ariane 5* rocket. To reduce the mass, each mirror segment is composed of beryllium, a lightweight, rigid, and thermally stable material, and thinly gold-coated to optimise the reflectivity at near-infrared wavelengths. An imaging camera, NIRCam, and a spectrograph, NIRSpec, operate over the wavelength range 1 to 5 microns. NIRSpec has a multi-slit capability, a first for a space-based instrument. Since multi-slit masks of the kind we used in LDSS (chapters 4 and 5) cannot be manufactured in space, the focal plane of NIRSpec comprises an array of 250,000 programmable "microshutters," each of which can be individually arranged to be "open" or "closed," thereby creating rectangular apertures for faint object spectroscopy. A third instrument, NIRISS, offers slitless spectroscopy at lower resolution for survey applications, and a fourth, MIRI, offers an imaging and spectroscopic capability from 5 to 27 microns. In order to maintain exquisite sensitivity, particularly at the longer infrared wavelengths, JWST must be kept at a temperature less than 50 degrees above absolute zero. This is achieved with a 15-by-21-metre, wafer-thin, five-layer sunshield, which protects the telescope from being warmed by solar and terrestrial radiation.

JWST was successfully launched from ESA's spaceport at Kourou, French Guiana, on Christmas Day 2021 (plate 56, *bottom*). Unlike HST, it is in a solar orbit at a special location called the second Lagrangian point (L2), about 1.5 million kilometres from the Earth in the direction away from the sun. Although JWST is further from the sun than the Earth, the significance of its special location (L2) is that the combined gravitational pull of the Earth and sun at that point equals that experienced by the Earth itself. This ensures that JWST and the Earth orbit the sun with the same period. It took JWST a month to reach this location, during which the primary and secondary mirrors and sunshield were carefully unfurled. This delicate operation was a time of great nervousness for many astronomers. Any technical problem could have jeopardised the entire $10 billion mission, since, unlike HST, it is not possible to visit or repair JWST.

It is often stated that JWST is the successor to HST, and in terms of its cost trajectory and political challenges over the years there are indeed similarities. However, the two facilities are quite complementary. HST is primarily an ultraviolet and optical observatory, whereas JWST is an infrared observatory. In some ways it is more appropriate to consider JWST as a spectacular successor to the Spitzer Space Telescope, both in terms of its enlarged aperture and significantly improved angular resolution. Since, even with its coarse angular resolution, Spitzer played such a significant role in studies of galaxies in the reionisation era, this augurs well for major progress in studies of cosmic dawn with JWST.

As we saw in chapter 6, Steidel's technique for locating distant galaxies via the telltale signature of ultraviolet hydrogen absorption requires a comparison of images taken in three colour filters adjacent in wavelength. The "dropout" signature in the shortest wavelength filter yields an approximate redshift (plate 34). Applying this technique with the limited infrared coverage offered by HST's camera, WFC3/IR, enables the detection of star-forming galaxies to a redshift limit of $z \sim 11$. Although Spitzer's imaging camera extended the coverage further into the infrared, its relatively poor sensitivity and low angular resolution meant it was ineffective in searches for higher redshift galaxies. Spitzer's role in studies of high-redshift galaxies has largely been in following up sources first detected with HST.

By comparison, NIRCam, the near-infrared camera on JWST, has the capability to select galaxies with high efficiency and HST-like angular resolution to redshifts beyond $z \sim 20$. However, at these longer infrared wavelengths the technique used so successfully to locate early galaxies with HST becomes less straightforward. Cool stars in the Milky Way have molecules in their atmospheres that produce deep absorption bands that can mimic the dropout signature of a high-redshift galaxy. This confusion is less of a problem at lower redshift, as the dropout signature lies in the optical region where these molecular bands are less prevalent. A further ambiguity of interpretation arises from the Balmer break, also due to hydrogen absorption but at optical wavelengths, discussed in chapter 9. A dropout signature due to ultraviolet hydrogen absorption at a redshift $z \sim 20$ might simply be a redshift $z \sim 6$ galaxy displaying a Balmer break. Again, this confusion is less problematic at lower redshift with HST because such a contaminant would be at lower redshift and probably be bright enough to be seen in other filters, which could be used to confirm this.

The first major observational project for JWST will be to undertake a census of star-forming galaxies beyond the $z \sim 11$ horizon accessible to HST. A key aim will be to resolve the long-standing dispute as to whether the abundance of galaxies declines rapidly beyond a redshift

$z \sim 8$ (discussed in chapter 8). Some theorists have postulated a long tail of early stellar systems out to $z \sim 20$ and beyond, which if found would suggest that age-dating the luminous sources at redshift $z \sim 9$ studied by Nicolas Laporte and my colleagues may not be the full story in pinpointing cosmic dawn.

Ultimately, spectroscopy will be needed to convince sceptics of the redshifts of sources claimed to be beyond a redshift $z \sim 11$; this will be very challenging, even with JWST. The spectrograph NIRSpec offers a remarkably wide wavelength coverage, enabling, in principle, access to familiar optical emission lines to redshift $z \sim 12$ and ultraviolet emission lines to redshift $z \sim 30$. However, the latter ultraviolet lines are usually faint, and, as discussed in chapter 9, ground-based telescopes have found it extremely challenging to routinely detect them in galaxies over the redshift range $7 < z < 11$. A widely discussed test for a first-generation galaxy, unpolluted by any nucleosynthesis, would be to recover a spectrum devoid of emission lines other than those of hydrogen and helium. NIRSpec is unlikely to detect the Lyman-alpha line of hydrogen, as it will not survive absorption in the surrounding intergalactic medium. Although other lines of helium, carbon, and oxygen can in principle be seen beyond $z \sim 20$, heroic exposures may be needed to detect them or place limits on their presence.

It's already clear from present studies that early galaxies are compact, with total stellar masses of only a few thousandths of that of the Milky Way. Nonetheless, they are likely to be forming stars much more prodigiously. Simulations suggest that the most massive stars in these primordial systems will explode within a few tens of millions of years after cosmic dawn, thereby immediately enriching the system with heavy elements. Thus the "window in time" where a young galaxy is completely devoid of heavy elements may be so short relative to the age of the universe at that time that such "pristine" galaxies will be exceedingly rare.

However, when cosmic dawn occurs it does also influence the nature of the cold gas in the intergalactic medium. Such cold hydrogen gas can be traced with radio telescopes using a spectrum line at a wavelength of 21 centimetres, which was first discovered in 1940 by Jan Oort, the

Dutch astronomer introduced in chapter 9, and later interpreted physically by his colleague Hendrik van de Hulst in 1944. Van de Hulst showed that this line originates in a very low (or "hyperfine") energy transition in the hydrogen atom that occurs when the orbiting electron flips from rotating one way to another. Radio astronomers have used this radio emission line to trace cold hydrogen across the entire Milky Way, as well as in external galaxies out to redshift $z \sim 5$. An enormous array of radio receivers is under construction in remote regions of Australia and South Africa called the Square Kilometre Array (SKA). SKA will have the sensitivity to map this 21-centimetre line during, and even before, the reionisation era. Since cold hydrogen gas pervades the dark ages, SKA has the capability of exploring the time before cosmic dawn. However, once cosmic dawn occurs, Lyman-alpha emission from the first stars can produce a detectable feature in the redshifted 21-centimetre signal seen in absorption against the cosmic microwave background (discussed in chapter 2). This effect was first predicted by Siegfried Wouthuysen (1916–1996), a Dutch physicist, and George Field, a Harvard astrophysicist. Although various groups are attempting to detect this signal ahead of completion of the SKA, it has not yet been seen convincingly. Even so, it may well become the most promising route to the ultimate determination of when cosmic dawn occurred.

After decades of pioneering observations with a sequence of ever-larger telescopes, will the Holy Grail of witnessing cosmic dawn be achieved with JWST? Certainly there are significant observational challenges in counting galaxies beyond $z \sim 11$, securing their spectroscopic redshifts, and studying their chemical composition. And, indeed, there may well be major surprises in store as we probe to less luminous sources than has been possible with current facilities. However, history encourages us to be optimistic. Astronomical observatories have generally outperformed the science goals described in their original motivating documents.

The Keck *Blue Book* written in January 1985 contains the science vision for a 10-metre telescope. Under the heading "How Do Galaxies Form and Evolve?," the following was written:

With the greatest of effort, present telescopes can image distant galaxies that carry us back in time two-thirds of the way to the Big Bang. These distant pictures show us what the Universe was like more than 10 billion years ago. . . . [U]nfortunately it is not enough: the pictures by themselves are ambiguous. They cannot tell us how far away a given galaxy is, and therefore cannot tell us the age of the Universe when the galaxy emitted the light we now image. The solution to this problem is spectra. . . . With spectra taken with the [Keck], we will be able to . . . follow the evolution of galaxies.[7]

In short, the spectroscopic horizon envisaged for the Keck Observatory was a lookback time of 10 billion years, corresponding to a redshift $z \sim 2$. In actuality, Keck has secured spectra for galaxies to redshift $z \sim 8.8$ and possibly to $z \sim 11$ (chapters 8 and 9), far outperforming its original purpose.

Such comparisons are similarly striking for the Hubble and Spitzer Space Telescopes. The report of the HST Working Group for Surveys (1985) discussed in chapter 6 was concerned largely with establishing that galaxies evolve. There is no mention of primordial systems or their role in ionising the intergalactic medium. Until Chuck Steidel's breakthrough applying the dropout technique to deep imaging data, the horizon for HST studies of distant galaxies was also a redshift just beyond $z \sim 1$. By the early 1990s, when what ultimately became the Spitzer Space Telescope was being planned, the goal of identifying primordial galaxies via infrared observations was finally recognised as important. In the science case for what was then called the Space InfraRed Telescope Facility (SIRTF), Mike Werner wrote, "Once a very young galaxy reaches an age of some tens of millions of years . . . much of its luminosity is produced by cool red giant stars" (emitting in the infrared).[8] It was

7. *The Design of the Keck Observatory and Telescope (Ten Meter Telescope)*, J. E. Nelson, T. S. Mast, and S. M. Faber (eds), Keck Observatory Report 90 (1985), pp1–4.

8. *The Scientific Case for SIRTF*, M. Werner (NASA, Jet Propulsion Laboratory, and California Institute of Technology 1990); see also Infrared Studies of Galaxies in Space, M. W. Werner and P.R.M. Eisenhardt, in *Airborne Astronomy Symposium on the Galactic Ecosystem: From Gas to Stars to Dust*, ASP Conference Series 73 (Astronomical Society of the Pacific 1995), pp169–76, quote at p3.

believed Spitzer had the sensitivity to search for such systems in the redshift range $z \sim 3$ to $z \sim 5$. However, at the time, the idea that Spitzer would play a pivotal role in characterising galaxies at redshifts $z \sim 9$ and beyond would have been regarded as fanciful. In a review talk celebrating 20 years of the Keck Observatory in 2013, I commented on these comparisons to an appreciative audience: "Unlike politicians, astronomers deliver much more than they predicted!"

ILLUSTRATION CREDITS

Figures

0.1. Timeline of book, Nicolas Laporte based upon redshift data compiled by Daniel Mortlock, Steve Warren and Nial Tanvir.

3.1. (*Left*) Sandage, A., "The Redshift-Distance Relation. II. The Hubble Diagram and its Scatter for First-Ranked Cluster Galaxies: A Formal Value For q_0", *Astrophysical Journal* 178, 1 (1972), Figure 11 (simplified); (*right*) Kristian, J., Sandage, A. & Westphal, J.A., "The Extension of the Hubble Diagram. II. New Redshifts and Photometry of Very Distant Galaxy Clusters: First Indication of a Deviation of the Hubble Diagram from a Straight Line", *Astrophysical Journal* 221, 383 (1978), Figure 3 (simplified). © American Astronomical Society. Adapted with permission.

4.1. (*Left, right*) Peterson, B.A., Ellis, R.S., Kibblewhite, E.J., Bridgeland, M.T., Hooley, T. & Horne, D., "Number Magnitude Counts of Faint Galaxies", *Astrophysical Journal* 233, L109 (1979), Figure 2. © American Astronomical Society. Reproduced with permission.

4.2. (*Left*) Shanks, T., Fong, R., Ellis, R.S. & MacGillivray, H.T., "Correlation Analyses of Deep Galaxy Samples—II. Wide Angle Surveys at the South Galactic Pole", *Monthly Notices of the Royal Astronomical Society* 192, 209 (1980), Figure 2; (*right*) Peterson, B.A., Ellis, R.S., Kibblewhite, E.J., Bridgeland, M.T., Hooley, T. & Horne, D., "Number Magnitude Counts of Faint Galaxies", *Astrophysical Journal* 233, L109 (1979), Figure 4. © Oxford University Press. © American Astronomical Society. Reproduced with permission.

4.3. (*Top*) Broadhurst, T.J., Ellis, R.S. & Shanks, T., "The Durham/Anglo-Australian Telescope Faint Galaxy Redshift Survey", *Monthly Notices of the Royal Astronomical Society* 235, 827 (1988), Figure 3 (portion); (*bottom*) Colless, M., Ellis, R.S., Taylor, K. & Hook, R.N., "The LDSS Deep Redshift Survey", *Monthly Notices of the Royal Astronomical Society* 244, 408 (1990), Figure 7 (portion). © Oxford University Press. Adapted with permission.

5.1. Observatorio de Roque de Los Muchachos, Julian Abrams.

5.2. Glazebrook, K., Ellis, R.S., Colless, M., Broadhurst, T., Allington-Smith, J. & Tanvir, N., "A Faint Galaxy Redshift Survey to B = 24", *Monthly Notices of the Royal Astronomical Society* 273, 157 (1995), Figure 2. © Oxford University Press. Reproduced with permission.

5.3. (*Top, bottom*) Crampton, D., Le Fèvre, O., Lilly, S.J. & Hammer, F., "The Canada-France Redshift Survey. V. Global Properties", *Astrophysical Journal* 455, 96 (1995), adapted from Figures 3 and 4. © American Astronomical Society. Reproduced with permission of David Crampton.

5.4. Ellis, R.S., Colless, M., Broadhurst, T., Heyl, J, & Glazebrook, K., "The Autofib Redshift Survey—I. Evolution of the Galaxy Luminosity Function", *Monthly Notices of the Royal Astronomical Society* 280, 235 (1996), Figure 10 (simplified). © Oxford University Press. Adapted with permission.

6.1. Advertisement in *Financial Times* 1977, author's copy.

6.2. (*Left*) Lyman Spitzer, Princeton Plasma Physics Laboratory, Denise Applewhite; (*right*) John Bahcall, Institute for Advanced Study, Randall Hagadorn.

6.3. (*Top, centre, bottom*) Couch, W.J., Ellis, R.S., Sharples, R.M. & Smail, I., "Morphological Studies of the Galaxy Populations in Distant 'Butcher-Oemler' Clusters with HST. I. AC 114 at $z = 0.31$ and Abell 370 at $z = 0.37$", *Astrophysical Journal* 430, 121 (1994), Figures 1 and 2 (portions). © American Astronomical Society. Adapted with permission.

6.4. (*Left, right*) Griffiths, R.E., Casertano, S., Ratnatunga, K.U., Neuschaefer, L.W., Ellis, R.S., Gilmore, G.F., Glazebrook, K., Santiago, B., Huchra, J.P., Windhorst, R.A., Pascarelle, S.M., Green, R.F., Illingworth, G.D., Koo, D.C. & Tyson, A.J., "The Morphology of Faint Galaxies in Medium Deep Survey Images using WFCP2", *Astrophysical Journal* 435, L19 (1994), Figures 1 and 3 (portions). © American Astronomical Society. Reproduced with permission.

6.5. Robert Williams, photograph by Stirling Colgate, permission from R. Williams.

6.6. Abraham, R.G., Tanvir, N.R., Santiago, B.X., Ellis, R.S., Glazebrook, K. & van den Bergh, S., "Galaxy Morphology to $I = 25$ in the Hubble Deep Field", *Monthly Notices of the Royal Astronomical Society* 279, L47 (1996), Figure 6. © Oxford University Press. Reproduced with permission.

8.1. Astronomers at Keck observatory, author's photograph.

8.2. (*Left, right*) Schenker, M., Stark, D.P., Ellis, R.S., Robertson, B.E., Dunlop, J.S., McClure, R.J., Kneib, J-P. & Richard, J., "Keck Spectroscopy of Faint $3 < z < 8$ Lyman Break Galaxies: Evidence for a Declining Fraction of Emission Line Sources in the Redshift Range $6 < z < 8$", *Astrophysical Journal* 744, 179 (2012), Figures 2, 5 (portion). © American Astronomical Society. Adapted with permission.

10.1. Delays in JWST launch, xkcd.com (free license).

Colour Plates

1. (*Left, centre*) *Out into Space* (Museum Press, 1954), author's copy; (*right*) author with Patrick Moore (2007), author's photograph.

2. (*Left, centre*) UCL Observatory, Ian Howarth, UCL; (*right*) solar eclipse, author's photograph (1971).

3. (*Left*) Herstmonceux Castle, Pixabay.com (free license); (*right*) Isaac Newton Telescope relocated on La Palma, Julian Abrams.

4. La Palma sunset, Julian Abrams.

5. (*Left*) Anglo-Australian Telescope control room, Julian Abrams; (*right*) Keck II control room, Julian Abrams.

6. (*Left*) Galileoscope in Hale control room, Julian Abrams; (*right*) library at Du Pont Telescope, Las Campanas Observatory, Julian Abrams.

7. (*Left*) Keck I primary mirror, Julian Abrams; (*centre, right*) Keck mirror segments, Julian Abrams.

8. (*Left*) Photograph of Andromeda spiral (cropped), Jay Bennett, *Smithsonian Magazine*; (*centre*) Hyper Suprime-Cam CCD detector, Satoshi Miyazaki, National Astronomical Observatory of Japan; (*right*) Hyper Suprime-Cam image of Andromeda spiral, Satoshi Miyazaki, National Astronomical Observatory of Japan.

9. Hubble Deep Field (portion), Zolt Levay (with permission from Robert Williams), STScI.

10. (*Top*) Redshift illustration, Mark Whittle, University of Virginia; (*bottom*) lookback time illustration, Nicolas Laporte, Cambridge University.

11. (*Left*) Telescope illustration, *Encyclopaedia for Children*, Odhams (1954), Future PLC; (*right*) Hale Telescope, Julian Abrams.

12. (*Left*) Allan Sandage and author (1990), author's photograph; (*right*) Hale Telescope, Julian Abrams.

13. (*Left*) Jim Gunn and the author (2010), author's photograph; (*right*) cluster 0024 + 16, M. James Jee, STScI.

14. Hale Telescope, Julian Abrams.

15. (*Left, right*) Anglo-Australian Telescope, Julian Abrams.

16. (*Left, right*) UK Schmidt Telescope, Julian Abrams.

17. (*Left*) Durham astronomers (1978), author's photograph; (*right*) Automated Plate Measuring Machine, Mike Irwin, Cambridge University.

18. (*Left, right*) Observing at the Anglo-Australian Telescope (1980s), author's photographs.

19. (*Left, right*) Fibre optic spectroscopy at the Anglo-Australian Telescope (1980s), author's photographs.

20. (*Left, right*) Observers at Coonabarabran airport (1980s), author's photographs.

21. (*Left, right*) Autofib positioner and components, author's photographs.

22. (*Left*) Multi-slit mask for LDSS, author's photograph; (*right*) Cassegrain cage of the Anglo-Australian Telescope, Julian Abrams.

23. (*Left, right*) The 2-degree field (2dF) fibre positioner, Julian Abrams.

24. Distribution of galaxies from the 2dF Galaxy Redshift Survey, 2dFGRS Team, Matthew Colless, Australian National University.

25. William Herschel Telescope, Julian Abrams.

26. (*Left*) Faint Object Spectrograph at the William Herschel Telescope, author's photograph; (*right*) design details of Faint Object Spectroscopy, author's drawings.

27. LDSS2 at the 6.5-metre Clay Telescope, Las Campanas, Julian Abrams.

28. (*Left*) Astronomers at Durham conference 1988, author's photograph; (*right*) astronomers at Leiden conference, early 2000s, author's photograph.

29. Hubble images of faint blue galaxies, Rogier Windhorst, STScI.

30. (*Left*) Cartoon from *Economist* July 14, 1990. © David Simonds; (*right*) Gemini South ground-breaking, author's photograph.

31. (*Left, right*) Gemini North telescope, Julian Abrams.

32. (*Top, bottom*) Gran Telescopio Canarias, Julian Abrams.

33. Spectroscopy in Hubble Deep Field, Mark Dickinson, Zolt Levay, STScI.

34. Dropout selection illustration, Chuck Steidel, Caltech.

35. Maunakea sunset, Julian Abrams.

36. (*Left*) Keck I telescope, Julian Abrams; (*right*) observing at Keck, author's photograph.

37. (*Left, right*) Astronomers involved in gravitational lensing, author's photographs.

38. The cluster Abell 2218, NASA, Andy Fruchter and the EarlyRelease Observations Team (STScI, ST-ECF).

39. The critical curve method, Ellis, R.S., Santos, M.R., Kneib, J-P. & Kuijken, K., "A Faint Star-Forming System Viewed Through the Lensing Cluster Abell 2218: First Light at $z = 5.6$?", *Astrophysical Journal* 560, L119 (2001), Figures 1 and 3 (modified). © American Astronomical Society. Adapted with permission.

40. (*Left*) Masanori Iye, with permission; (*right*) Iye, M., Ota, K., Kashikawa, N., Furasawa, H., Hashimoto, T., Hattori, T., Matsuda, Y., Morokuma, T., Ouchi, M. & Shimasaku, K., "A Galaxy at a Redshift $z = 6.96$", Nature, 443, 186 (2006), Fig 1, 2 (modified). Reproduced with permission by Springer Nature.

41. Birthday of Sir Fred Hoyle 1995, Institute of Astronomy, Cambridge University.

42. Illustration of cosmic reionisation, Brant Robertson, University of California Santa Cruz.

43. (*Top*) Planck observations of microwave background, ESA; (*bottom*) illustration of electron scattering by ionised intergalactic medium, ESA.

44. (*Left*) Astronomers at the Royal Observatory, Edinburgh, 2021, Brant Robertson; (*right*) Ultra Deep Field, Richard Ellis, STScI.

45. (*Top, bottom*), Robertson, B.E., Ellis, R.S., Furlanetto, S.R. & Dunlop, J.S., "Cosmic Reionization and Early Star-Forming Galaxies: A Joint Analysis of Constraints from Planck and the Hubble Space Telescope", *Astrophysical Journal* 802, L19 (2015), Figures 1 and 2 (modified). © American Astronomical Society. Adapted with permission.

46. Cerro Tololo Inter-American Observatory, Julian Abrams.

47. (*Top pair*) Cerro Paranal and ESO's Very Large Telescope, Julian Abrams; (*bottom left*) unit telescope 4 at the Very Large Telescope, Julian Abrams; (*bottom right*) ESO residence at Cerro Paranal, Julian Abrams.

48. Cerro Armazones, author's photograph.

49. Spitzer excess galaxies, their spectral energy distributions and spectra, assembled by Guido Roberts-Borsani.

50. (*Top*) Pirzkal, N., Coe, D., Frye, B.L., Brammer, G., Moustakas, J., Rothenberg, B., Broadhurst, T.J., Bouwens, R., Bradley, L., van der Wel, A., Kelson, D.D., Donahue, M., Zitrin, A., Moustakas, L. & Barker, E., "Not in our Backyard: Spectroscopic Support for the CLASH z = 11 Candidate MACS 0647-JD", *Astrophysical Journal* 804, 11 (2015), Figure 3; (*bottom*) Zitrin, A., Zheng, W., Broadhurst, T., Moustakas, J., Lam, D., Shu, X., Huang, X., Diego, J.M, Ford, H., Lim, J., Bauer, F.E., Infante, L., Kelson, D.D. & Molino, A., "A Geometrically Supported z ~ 10 Candidate Multiply-Imaged by the Hubble Frontier Fields Cluster A2744", *Astrophysical Journal* 793, L12 (2014), Figure 1 (portion). © American Astronomical Society. Reproduced with permission of Nor Pirzkal and Adi Zitrin.

51. (*Left*) Pascal Oesch, by permission; (*centre, right*) Oesch, P.A., Brammer, G., van Dokkum, P.G., Illingworth, G.D., Bouwens, R.J., Labbé, I., Franx, M., Momcheva, I., Ashby, M.L.N., Fazio, G.G., Gonzalez, V., Holden, B., Magee, D., Skelton, R.E., Smit, R., Spitler, L.R., Trenti, M. & Willner, S.P., "A Remarkably Luminous Galaxy at z = 11.1 Measured with Hubble Space Telescope Grism Spectroscopy", *Astrophysical Journal* 819, 121 (2016), Figures 5 and 6 (portions). © American Astronomical Society. Reproduced with permission of Pascal Oesch.

52. Astronomers at Cerro Paranal, author's photograph.

53. Hashimoto, T., Laporte, N., Matawari, K., Ellis, R.S., Inoue, A.K., Zackrisson, E., Roberts-Borsani, G., Zheng, W., Tamura, Y., Bauer, F.E., Fletcher, T., Harikane, Y., Hatsukade, B., Hayatsu, N.H., Matsuda, Y., Matsuo, H., Okamoto, T., Ouchi, M., Pelló, R., Rydberg, C-E., Shimizu, I., Taniguchi, Y., Umehata, H. & Yoshida, N., Nature 557, 392 (2018) Figures 1 and 2, Methods Figure 2 (portions). Credit: Takuya Hashimoto, Springer Nature.

54. Illustration of cosmic dawn, Nicolas Laporte, Cambridge University.

55. Cosmic dawn simulation, Harley Katz, Oxford University.

56. (*Top*) The James Webb Space Telescope at Northrop Grumman, El Segundo, CA, 2019, NASA; (*bottom*) view of James Webb after launch (ESA).

INDEX

Numbers in *italics* refer to pages with illustrations.